China Academic Library

More information about this series at http://www.springer.com/series/11562

Shuyuan Lu

The Ecological Era and Classical Chinese Naturalism

A Case Study of Tao Yuanming

外语教学与研究出版社
FOREIGN LANGUAGE TEACHING AND RESEARCH PRESS

Springer

Shuyuan Lu
School of Humanity
Soochow University
Suzhou
China

Translated by Meng Xiangchun

ISSN 2195-1853 ISSN 2195-1861 (electronic)
China Academic Library
ISBN 978-981-10-1782-7 ISBN 978-981-10-1784-1 (eBook)
DOI 10.1007/978-981-10-1784-1

Jointly published with Foreign Language Teaching and Research Publishing Co., Ltd

Library of Congress Control Number: 2016954550

Printed on acid-free paper

This Springer imprint is published by Springer Nature
The registered company is Springer Nature Singapore Pte Ltd.
The registered company address is: 152 Beach Road, #22-06/08 Gateway East, Singapore 189721, Singapore

Dedicated to Tao Yuanming and the world's eco-movement

Foreword

Returning

Tao Yuanming represents the most perfectly harmonious and well-rounded character in the entire Chinese literary tradition. There is a simplicity in his life, as well as in his style, which is awe-inspiring and a constant reproach to more brilliant and more sophisticated natures. And he stands, today, as a perfect example of the true lover of life, because in him the rebellion against worldly desires did not lead him to attempt a total escape, but has reached a harmony with the life of the senses.

—Lin Yutang in *The Importance Of Living* (1937)

Tao Yuanming (365–427), one of China's most celebrated poets, presents us with a vision not of escape from this world, but of return to it. This might be described as "living in the presence of things." Dwelling in his country home, peering through the mists at dawn, soaking in the rising sun. Hearing bird song like incense rising; delighting in monkey calls like children at play.

With these feelings aroused in the presence of vibrant life how can one not sink into immense and lasting gratitude? How can one not become a person of reflection and contemplation? How can compassion not flow forth?

Lu Shuyuan explores these possibilities in his splendid new book, recognizing the potential and limits of Tao Yuanming's voice in our world. He wonders, sixteen centuries later, what is it that Tao offers to us in our hyper-industrialized world? Perhaps it is a spare yet penetrating vision of the luminous quality of nature that illuminates our own mind and heart. This returning gives us a sense of how we belong.

As we face the rising sun, Tao shows us another dimension that comes flooding across us and into our heart. This gives rise to a sense of awe and reverence. We are engulfed in a depth of beauty without end, a presence of things to one another.

This sensibility is what Tao Yunaming longed to cultivate—a spirit attuned to the heart of matter. He left public office disillusioned by the power struggles involved in politics:

Long have I been imprisoned in the cage;
Now back to Nature I return again.

From "Returning to my Farm Young"

He sought refuge in a country setting—searching for serenity in nature. His soul yearned for that connection that can be felt and intuited more than spoken. Through his poetry he brought forth penetrating glimpses into this connectedness—the return to things.

In the flight of a bird, in the fullness of autumn chrysanthemums, in the sound of a river, in the expanse of an open field, in familial pleasures he found solace—indeed joy. Those rare gifts of peacefulness and ecstatic joy are what Tao celebrated in his poetry, often while drinking wine. And these sensibilities pass through the ages across sixteen centuries to speak to us still.

Tao Yuanming is more than simply a romantic poet or another nature writer. He is a poet beyond conventional labels or descriptions. He is a seer into the soul of things, capturing the inner vibrating life of nature in its simplest movements and changing forms. He touches lightly, yet deftly, on these vibrations so that we sense the pulsating energy—the *qi* of things arises to meet us in his words and in his images that linger in the mind. Rarely have form and energy been mingled so powerfully in the elixir of the imagination. Rarely have the concealed and the revealed been so skillfully interwoven.

Qi arises in his poetry to meet us with subtlety like the qiyung in Sung landscape painting. This qiyung—spirit resonance—is palpable in Tao's poems and in landscape paintings. We can taste the energy, sense the beauty, and drink in the pulsating life. Mists arise in the craggy mountains where the human figure is slight yet fits in.

How can we moderns miss this in our rush to manipulate nature—making and manufacturing things endlessly, often without a sense of consequences? How can we have reached such extremes beyond the limits of natural growth? The story is long and complex, but now well known. The industrial revolution has reached its limits. We are shutting down ecosystems at a rapid rate, and thousands of species are going extinct. Waste is accumulating and toxic pollution has penetrated our bodies and Earth's body.

There seems no end in sight to the levels of destruction. Has modernity reached an impasse such that we must make a turn. But where? The suggestion of a "return to nature" seems unrealistic, quaint, and rather out of fashion. And yet isn't there a way to find our way back to hearth and home? Isn't there a way to discover again our place in the midst of nature's creative unfolding? Isn't there a way to realize we are part of an immense journey of life?

This is the way of Tao Yuanming as he offers a path through the thicket. It is not with a backward glance toward wilderness renewal or bucolic rest. It is rather with a sense that one can dwell in the midst of the endless comings and goings and still cultivate serenity. One of his most famous poems captures this stance of detached entanglement with the world.

Drinking Wine
I made my home amidst this human bustle,
Yet I hear no clamor from the carts and horse.
My friend, you ask me how this can be so?

A distant heart will tend towards like places.
From the eastern hedge, I pluck chrysanthemum flowers,
And idly look towards the southern hills.
The mountain air is beautiful day and night,
The birds fly back to roost with one another.
I know that this must have some deeper meaning,
I try to explain, but cannot find the words.

Tao Yuaniming offers a way of being—a penetrating engagement with the life energies that will sustain humans. Intimacy and distance are mingling here. One's mind and heart is renewed daily. Such self-cultivation (Daoist, Buddhist, or Confucian), he intuits, is what grounds humans for the work to be done. Amid movement he knew the power of detachment:

One world. Though the lives we lead are different,
In courts of power or laboring in the market,
These I know are more than empty words:
Our life's a play of light and shade,
Returning at last to the Void.

His sense of the play of light and shade was certainly present in his poem "Substance, Shadow, and Spirit," which featured a conversation between these entities representing Daoist, Confucian, and Buddhist positions.

"Peach-blossom Springs" is one of his most iconic pieces. There he presents a kind of Shangri-la, a village where people live in peace and harmony. But when the fisherman who visits there tries to find it again he is unsuccessful. The message is that such a utopian community is illusive to realize and difficult to sustain. Yet the image of Peach-blossom Springs has endued through the ages as an ideal world in the East Asian imagination and beyond.

Sixteen centuries have passed since Tao Yuanming lived. Our planet is a different place. Our period is now called the Anthropocene because of our human imprint everywhere. Yet Tao's remarkable insights into the mysterious resonance of nature provide a guiding thread. This thread weaves us back in to the web of life. And this web is filled with both creativity and destruction that often clouds our mind with blindness and sorrow.

Yes, Tao Yuanming knew the traps of ignorance and the vagaries of life's tragedies. Yet his insight into nature's radiance and human belonging provides restorative joy now and in the centuries to come. These poetic offerings will endure and help us to do likewise. For more than ever before, we are seeking insight into being part of a larger universe that will hold us steady amid the destruction of ecosystems and disintegration of societies.

For Tao Yuanming, this is expressed with utter simplicity and comprehensive inclusion:

"At a single glance I survey the whole Universe.
He will never be happy, whom such pleasures fail to please!"

That same sense of belonging to the universe is what Albert Einstein sought when he reflected: "A human being is a part of a whole, called by us 'universe,' a part limited in time and space. He experiences himself, his thoughts and feelings as something separated from the rest … a kind of optical delusion of his consciousness. This delusion is a kind of prison for us, restricting us to our personal desires and to affection for a few persons nearest to us. Our task must be to free ourselves from this prison by widening our circle of compassion to embrace all living creatures and the whole of nature in its beauty."

This book will open up Tao Yuanming to readers eager to hear his poetry, drink wine with him, and listen to his music played across the ages. Through Tao, we may return to the universe.

Mary Evelyn Tucker
Yale University

Preface

Born in Chaisang, Xunyang, Jiangxi province, at the end of the Eastern Jin Dynasty, Tao Yuanming (365–427) was, is, and perhaps will always be, a great poet and philosopher, and a symbol of classical Chinese naturalism.

Although there are just over one hundred of his works available today, Tao[1] has a holy status in the history of Chinese literature and is accredited "the poet of poets" because he was naturally inclined toward Nature[2] and would find himself chanting Nature since Nature was part of his life. While living to the rhythm of Nature, he was bestowed with maximum freedom, thus becoming a fundamentally beautiful, though simple, epitome of poetic dwelling on the earth. His complex of the *tianyuan*[3], "returning to life's essentials," and "aspiration to the Peach-blossom Springs" have become spiritual symbols of traditional agricultural China and been built into the collective unconscious of the Chinese.

Tao's works appeal to us as "Nature" itself does. His greatness lies in his integration with and immersion in "Nature." His destiny, therefore, is interwoven with that of "Nature."

Unfortunately, his glory has been dramatically eclipsed by the ever-rumbling industrialization, commercialization, urbanization, and the deterioration

---

[1]Tao is used to refer to Tao Yuanming for the sake of convenience. Coincidentally, what is known as *Tao*, or the way, as in Taoism represented by Lao Tzu and Chuang Zi, is spelled the same as Tao as a surname. Therefore, to avoid confusion, *Tao* as a philosophical concept will be spelled as Dao and the philosophy of Dao as Daoism. Translator's note.

[2]Nature with a capitalized N when in the middle of a sentence refers to Nature in the Chinese sense of the word, meaning the natural world and the so-of-itself in most cases. "Man and nature" is used hereinafter for discussions of quintessentially Western scholars' contemplation on the relationship between man and the natural world; "Nature and man" is used for most Chinese scholars' discussions on the relation in question, since, in classical Chinese philosophy and discourse, Nature almost always precedes man. Translator's note.

[3]Tianyuan, literally translated as fields and gardens, refers to the country of rustic charm, Arcadian simplicity, and contentment. Tianyuan poetry, or farmstead poetry, is similar to English pastoral poetry in terms of aesthetic experience and idealized imagery, though *tianyuan* poetry is not about herding.

of “Nature” itself. As Nature suffers, Tao’s literary spirit also wanes. The early twenty-first century witnessed Tao’s death again, following his true death approximately 1600 years ago. Consequently, Tao has been reduced to a feeble and elusive specter.

Serge Moscovici (1925–), a contemporary French eco-critic, declares that “Nature” would be resurrected as a specter in the human world by the global ecological movement:

Nature for a time disappeared from our vision,
Hiding behind a word we rarely use.
Yet nature has resurrected,
Reminding us that it is still there,
And that lives all rely on it.
The reapparition of nature is no radiant presence,
But like a ghost returning to the human world,
And the naturalist and ecologist movement is becoming
A lingering specter on the Euro-American lands. (87)

He Lin (1902–1992), a Chinese philosopher, writes about Tao Yuanming equally poetically:

Nature, once internalized, radiates in our souls.
The tranquil Southern Mount as depicted by Tao Yuanming,
The Peach-blossom Springs as pursued by the fisherman of Wuling,
And all the landscapes in poems and paintings,
May be viewed as
Radiating Nature in souls. (112)

I attempt to interpret Tao Yuanming in a postmodern eco-critical context from multi-disciplinary and cross-cultural perspectives. It is my wish, and perhaps yours too, that the specter of Tao may rekindle a light for humanity in an era of suffocating smog and filthy water, a light that will help evoke a completely intact “Nature” in our minds and souls. That very Nature, in turn, is the real fountainhead of a truly beautiful human existence.

Suzhou, China
June 2012

Shuyuan Lu

Acknowledgements

Great poets, no matter what nationalities or races they belong to, are, in a sense, philosophers about man, life, and nature. Hölderlin and Heidegger of Europe prescribed "poetic dwelling on earth" for modern people, whereas Tao Yuanming, as far back as 1600 years ago, already became an epitome of "poetic dwelling" as demonstrated by his lived experiences and writings. Since the advent of the industrial age, human beings, running counter to Mother Nature, have been alienated from Nature. Consequently, the wrath of ecological crises is ubiquitously and more keenly felt. Classical Chinese naturalism that Tao Yuanming embodies may, if not will, offer some revelations for people today in their attempt to come out of the ecological impasse. When I decided to write this book about ten years ago, I began to explore Tao Yuanming's "international soulmates" in the hope of situating Tao in a Western context juxtaposed with the Chinese cultural context, and of dedicating classical oriental wisdom about human existence to the ever-growing ecological movement of the world.

After its publication in China in 2012, the book, originally titled *The Specter of Tao Yuanming*, received tremendous critical attention of many Chinese scholars and strong interest of some international critics. It received the prestigious "Luxun Literary Award" of 2014. I am sure that the publication of the English edition of the book by the time-honored and widely respected Springer will contribute to the visibility and reception of Tao Yuanming in the Western critical eye. For this reason, my sincere thanks go first and foremost to Springer. My thanks also go to Prof. Yue Daiyun of Peking University, the chief editor of *Cross-cultural Dialogue*, for her strong recommendation of the book on many occasions, and to Mr. Wu Hao and Ms. Yi Lu of the Foreign Language Teaching and Research Press for their liaison which has made the publication of this book possible.

I am indebted to Dr. Meng Xiangchun of the School of Foreign Languages, Soochow University, for this translation, or translation-rewriting, of the book, and to his collaborator Carl Ballantine, an experienced proofreader well versed in

Chinese language and culture. During the translation-rewriting of this book over a course of up to 2 years, Dr. Meng, a trustworthy "ferryman" in the East–West dialogue, touched me with his knowledge, language competence, insights, discretion, and modesty.

I am grateful to my long-standing friend Prof. Wang Zhihe, director of the China Project of the Center for Process Studies and chief editor of *Process Studies in China*, and his wife Fan Meiyun, for their generous support; to John B. Cobb, Jr., a member of the American Academy of Arts and Sciences for his understanding, trust, and recommendation; and to Prof. Mary Evelyn Tucker of Yale University for her poetically insightful Foreword to this book.

My thanks also go to Prof. Yu Mouchang, Prof. Zeng Fanren, Prof. Wu Bingjie, Prof. Li Xiaojiang, Prof. Chen Xiaoming, Prof. Wang Nuo, Prof. Cheng Xiangzhan, Dr. Wei Qingqi, Dr. Hu Zhihong, Dr. Liu Bei, and many other scholars for their eco-critical engagement and inputs. Prof. Wang Xianpei, in particular, has been a constant encouragement and inspiration for my research. On those lingeringly rainy autumn days several years ago, he kept me company exploring Tao Yuanming's hometown and personal traces. Prof. Peter I-min Huang of Tamkang University and Prof. Lin Baqian of Soochow University invited me to their universities in Taiwan to discuss the contemporary significance of Tao Yuanming. I would like to thank Prof. Arnold Berleant, a renowned environmental aesthetician, former secretary-general and past president of the International Association of Aesthetics, Prof. Scott Slovik, a reputable eco-critic, and environmental aesthetician Yrjö Sepänmaa for their inputs at my eco-critical center.

I am deeply indebted to my wife Zhang Ping, a professor at the Faculty of Music, Soochow University. She is, as always, a kindly presence, a constant inspiration, and a reassuring support for me and my research project.

No decent research or academic writing is possible without the foundation laid by our predecessors. I would like to thank more than one hundred scholars at home and abroad I cite in this book for the possibility of weaving in a mutually illuminating and mutually completing way their quotes and my own writing into an open-ended textual network or an interesting Gestalt before they are eventually integrated into a vaster sea of texts.

Suzhou, China
June 2016

Shuyuan Lu

Contents

Abbreviations

II	A Complete Collection of the Three-Pine Study (II)
V	A Complete Collection of the Three-Pine Study (V)
BLD	The Bottom Line of Derrida
BLR	Beyond Left and Right
CB	The Continuity of Being: The Chinese Version of Nature
CC	The Confessions and Correspondence
CE	Civilization and Ethics
CM	The Consequences of Mordernity
CWJ	The Complete Works of Jin Yuelin
DI	Dialogue and Innovation
DPESC	Discourse on Political Economy and the Social Contract
E	Émile
EHP	Elucidiations of Hölderlin's poetry
FE	Forgetting and Evocation
HLT	A History of Literary Thought: From 220 to 589
MD	A Mock Dialogue with Derrida
MM	Metaphysics and the Mindset of the Wei and Jin Scholars
MPN	Man's Place in Nature
PC	The Philosophy of Civilization
RSW	The Reveries of the Solitary Walker
SHCH	A Short History of Chinese Philosophy
SJR	Shang Jie on Rousseau
SSWL	Self-Selected Works of Liang Shuming
SWS	Selected Works of Max Scheler
TNM	Tao, Nature and Man

Translator's Note

Translation as a Cross-cultural Critique in the Ecological Era

In this ecological era, translation, in addition to its role of intermediation, is essentially a cross-cultural critique of human existence, of human thought, and of their pursuits, prospects, and potentials. It may help create a special, though not necessarily always duly respected, mode of scholarship that involves many rather than one, though it does face a plethora of challenges. Such endeavor may inform and inspire readers in the target language by providing something heterogeneous or different since they, like human beings in general, tend to take for granted their own culture and everything about, by, or beneath it.

The foremost task of a cross-cultural critic, which happens to be me in this case, is to make accessible to Western readers the classical Chinese concepts found in this book, philosophical or poetic, which, for Stephen Owen (1946–), are characterized by imagism, impressionism, and vagueness. First, I have to examine a particular Chinese concept and then judge critically all possible Western concepts that may serve as "equivalents." In most cases, translation, whose nature is rewriting, is a critique-based compromise. For instance, *ziran* (自然), usually translated as "nature," is, in fact, "the so-of-itself," of which the natural world is only part. To convey this difference, I use Nature with a capitalized N to refer to *ziran* and nature for the natural world in the Western sense of the word. *Yuzhou* (宇宙), rendered as the universe, is a temporal–spatial concept, while "universe" is just a spatial concept. *Yijing* (意境), an ethereal Chinese aesthetic category, means the evoked poetic realm featuring a subtle unconscious blending of the aesthetic subject's idea/thought, mood, and the imagery out there. Only when there is the oneness of the mind (reason), the heart (emotions, feelings, mood, etc.) and the scene, real or imagined, is *Yijing* possible. The mutual generation of reality and imagination and the infusion of the subject's mood and thought into the imagery out there can arouse an aesthetic experience in the viewer or reader as if he or she were physically situated in the world created by the piece of work he or she is appreciating. Stephen Owen has translated *yijing* into "artistic/poetic conception" or sometimes "the divine realm," whereas others have rendered it into "the frame of mind" (Liang Shiqiu), or the poetic world, the poetic space, the artistic reality, etc. I must view these renditions critically and try to find out what kind of aesthetic experience they may generate in the West before I decide to create the expression "ideo-mood-imagery, or the evoked poetic realm." Similarly, *wei* (味) is rendered into "taste" in most cases in this book, but I am fully aware that as a classical Chinese aesthetic concept, it points to the quality of a piece of art work in the Chinese context instead of an ability in the Kantian sense of the word, which is sensitivity to the aesthetic properties of objects and aesthetic intuition and response which enable the subject to judge, for instance, what is elegant, beautiful, or sublime. In some extreme cases, I have to read a few books or at least several articles before I may have a reasonable cross-cultural critique of even a single Chinese concept. For instance, to translate *shen* (神, literally meaning "spirit," or the soul), I have to turn to Western philosophy, for instance, Plato's tripartite theory about the soul, Marx's and Derrida's theory about specters, and the perception of ghosts in the Christian tradition. Technically, the book *per se* is a result of a cross-cultural critique on the part of the author in the first place and the translator in the second. To make classical Chinese concepts understandable without being too complicated or lengthy, I, while trying to preserve their Chineseness if I can, usually choose to view them through a Western lens and, as a result of that viewing, I may be able to re-present them with transformed Western ideas when necessary, as shown by such expressions as "the mysterious female reminiscent of Gaia," "all natural beings as the so-of-itself being in a perfect state of being," and "the Peach-blossom Springs as an Arcadia, or a Shangri-la, or a land of promise". In most cases, I juxtapose inwardly what is Chinese and what is Western before synthesizing them in a new holistic representation. This practice can be described as thinking critical and speaking glocal, which, in today's context, has some sort of ecological significance.

The author of the book, who perfectly understands the nature of my work, has given me much leeway to re-present in the target language only contents that I assume may be meaningful or significant for target readers. I can and must use discretion in deciding what to be conveyed in which way. This book is about philosophy and poetry—and very Chinese philosophy and poetry, among other things. "Suggestiveness, not articulateness, is the ideal of all Chinese art, whether it be poetry, painting or anything else" (Feng You-lan 1947), and such ideal is not without its philosophical background. Chuang-tzu (born approximately in 369 BC) argues that "words are for holding ideas, but when one has got the idea, one need no longer think about the words" (*Chuang-tzu* Chap. 26). Chinese philosophy and poetry are highly elusive, and their suggestiveness makes my work even more difficult. In a sense, philosophy and poetry are what get lost in translation, but it is also true that they are what get created in translation. This seems to be the added value of translation as a cross-cultural critique. I will have tremendous satisfaction if, through my translation, readers can discover a Tao Yuanming strange yet familiar to them, informing their own existence in nature.

Texts are extra-texts. I interpret Lu Shuyuan's text as the crystallization of his lived experiences and his contemplation of humanity's reflection on poetic dwelling as represented by Tao Yuanming, to which people today have good reason to aspire. In addition to the critique of the text, I also had a cross-cultural critique of Lu not simply as the author of the book, but also as a philosophically and poetically-inclined man who is committed to finding for human beings a way out of the ecological impasse. In order to better re-present his thought, I communicated with him on various occasions, and I read his earlier writings and the works of some Western eco-critics who I believe might be Lu's "allies" across time and space. Unlike Tao Yuanming or Wendell Berry, who returned to the country for a more natural existence, Lu still lives in the city. However, he lives in the city without the city: He uses the modern conveniences the city offers to write in order to impact people, but his mind and heart dwell poetically in his own "Peach-blossom Springs" somewhere else. For me, both Lu and his book are about "return" in the Anthropocene, a spiritual return to the natural, true, and beautiful human existence.

Hu Shi (1891–1962), former president of Peking University, argues that if a civilization "makes the fullest use of human ingenuity and intelligence in search of truth to control nature and translate matter to serve mankind, to liberate the human spirit from ignorance, superstition, and slavery to the elements, and to reform social institutions for the benefit of the greatest number—such a civilization is highly idealistic and spiritual. This civilization will continue to grow and improve itself, but its future growth and improvement will not be brought about by returning to the spiritualistic ideals of the East." First of all, I doubt human beings' ability to "control" nature. Secondly, if the "future growth and improvement" means people's greater well-being, spiritually or in practical terms, then I think Lu's answer, which is different than Hu's, has more legitimacy and ecological significance.

Some scholars seem to have overemphasized differences between the West and the East; hence, there have emerged "orientalism," "occidentalism," and an impulse to perpetuate the "otherness" of cultures different from their own. I am aware that the much-talk-about discrepancies between Chinese culture and Western cultures (I do not endorse the idea of Western culture since it is no homogeneous entity) such as holism vs. analysis, collectivism versus individualism, intuition versus logic, femininity versus masculinity, atheism versus religiosity, imagism versus positivism, vagueness versus accuracy, introversion versus extraversion, and induction versus deduction, past-orientedness versus future-orientedness, to name but a few, are nothing but dangerously convenient over-generalizations or stereotypes. In fact, humans are humans: defined by their shared humanity and hewn from the same material, they are basically the same, bound together by the same sense of joy, achievement, and suffering, especially in a world bristling with problems including the still mounting eco-crisis. It is important to highlight differences between cultures, but it is not less significant to explore universals shared by all cultures in our attempt to shape our world as an ecosystem of common destiny. I am very happy to see that Prof. Lu not only elucidates Tao Yuanming's unique identity as an avatar of classical oriental romantic naturalism, but also presents to us Tao's Western naturalist "soul mates." Regarding the shaping of our common ecological future, establishing "unity in diversity" is as important as preserving "diversity in unity."

Professor Lu describes the translator as a "ferryman" between cultures, which is an accurate observation. Different cultures may have equally effective though vastly different insights into or ways out of the ecological impasse; therefore, translation as a cross-cultural critique may help retrieve or re-establish in this earthly world of ours a natural and poetic dwelling. Perhaps it is time that we resumed the building of Babel though we do speak different languages.

Meng Xiangchun
Soochow University

Chapter 1
Tao Yuanming and the Meta-question of Humanity

Europeans used to consider "man and God" as a meta-question that human beings had to solve, and people in a near-ecological era, however, tend to view "man and nature" as an inevitable meta-question of humanity. In traditional Chinese culture, "Nature and man" means "Heaven and man," "the interaction between man and Heaven," or "the oneness of man and Heaven." Technically, Heaven in this case denotes both Nature and the Chinese version of God. "Nature and man" in the Chinese sense, therefore, seems to be more inclusive and complex than "man and God" in quintessentially Western discourse.

In *Tao Yuanming* written in his twilight years, Liang Qichao (1873–1929) praises Tao as a supreme example of the harmony between "the appeals of Nature" and "the enchantments of life" (9). A reinterpretation of Tao may offer an oriental solution to the issue of "Nature and man" as the real meta-question and give people a ray of hope for readdressing the ever-worsening ecological crisis of the world.

The prefix "meta" as in "meta-question" means "higher or beyond," and "meta-question," in turn, means the "essential question" that transcends concrete or specific questions, often in the form of an "ultimate question" that is beyond abstraction, formalization, or logical operation. Meta is translated into *yuan* (元) in Chinese, meaning originally the mind or the root of life, and, derivationally, primordial, primary, foremost, fundamental and significant. Its combinations such as life-force (*yuanqi*), the mandate of Heaven (*yuanming*), and the creation of the universe (*yuanhua*) are often related in one way or another to the cosmological entities and/or their operations. Besides, *yuan* also means "profound, mysterious, or metaphysical."

I intend to infuse into the Western term "meta-question" with some typically Chinese images or auras. Rather than a result of abstraction, formalization, or logical operation, a meta-question in this book means the very "primordial, primary, and significant" question about humanity. It precedes all other questions temporally and covers or contains them spatially. Therefore, it defines the nature of, and solutions to, all other questions. It will exist as long as other questions do.

S. Lu, *The Ecological Era and Classical Chinese Naturalism*,
China Academic Library, DOI 10.1007/978-981-10-1784-1_1

Undoubtedly, such a meta-question is none other than the issue of "Nature and man." The way people address this issue not only determines the nature of human society but also the zeitgeist and even the physiological makeup of mankind in a given era. Unfortunately, though there have been sporadic warnings from thinkers and saints, human beings have, for a long time, either turned a deaf ear or responded wrongly to the crucial meta-question, thus having caused the devastating eco-crisis today.

How come Tao has made it to a holy position in the Pantheon of Chinese literature and is dubbed "the poet of poets," though his works available today are far from being plentiful? The answer lies in his lived solution to the issue of "Nature and man."

Naturally inclined, Tao paid homage to Nature and committed himself to it. More often than not, he would find himself chanting Nature since Nature was part of his inspiration, vitality, and life. As he lived to the rhythm of Nature, he was bestowed with maximum freedom, thus becoming a fundamentally beautiful though simple epitome of human existence, and of dwelling poetically on the earth. Hopefully, my interpretation of Tao can offer new possibilities in exploring the issue of "Nature and man."

1.1 Tao as an Oriental Prophet of the Meta-question of Humanity

As a philosophical masterpiece of ancient China for China-West cultural dialogues, *Lao Tzu (Dao De Jing)*, which addresses issues of the natural world and the human world, is an ancient Chinese thinker's esoterically subtle interpretation of "Nature and man" as the meta-question.

In his 1943 lecture in the USA, Jin Yuelin (1895–1984), rooted and versed in classical Chinese philosophy, elaborates on the issue of "Nature and man" in the hope of introducing healthy traditional Chinese cultural spirit into a highly developed industrial society where reign the notions of human rights, humanism, the rule of law, and a science-worshipping cultural–intellectual community. Truly laudable is Jin's aspiration and effort.

In the lecture, Jin summarizes the backbone of Chinese philosophy into "the unification of nature and man" (in Jin's term) or "the oneness of Nature and man." He emphasizes that Nature in the Chinese context is not merely a cognitive object, but rather a subject of worship, an important source of faith, a carrier of human feelings and emotions, and a complex conceptual image (Jin Yuelin "TNM" 174–175). He believes that *tian*, or Heaven, as the Chinese see it, is not "mere nature" in the Western sense of the word, but a complex and inclusive concept, reminiscing of the sanctified or consecrated Nature, which, in turn, is akin to "Nature's God" (175).

Jin points out that in traditional Chinese culture, the view of life and that of Nature are interconnected in that one's attitude toward Nature also defines one's attitude toward life. Interestingly, Jin identifies three types of "vistas" (As Jin puts it, "The word vista means 'firstly a view or prospect commonly through or along an avenue as between rows of trees, or secondly a mental view or prospect extending over a series of events.' We shall keep the view or prospect of the meaning and discard the tree or the event part.") namely "the simple vista," "the heroic vista," and "the sagely vista." The "unification of Nature and man" means the harmony between Nature and man. The relationship between Nature and man involves the individual's values and way of life as well as many other issues that human beings face, including social ideals, social order, ethos, aesthetic inclinations, or paradigms. In a word, Jin offers a near-perfect Chinese-style interpretation to the meta-question of "Nature and man" in a Western context.

Lao Tzu argues that "[*D*]*ao* bore them and the power of '*Dao*' reared them, made them grow, fostered them, harbored them, and brewed for them." (109) "Therefore, of the ten thousand things, there is not one that does not worship *Tao* and do homage to its 'power.'" (109) These quotes point to the harmonious interaction between the Way of Nature and that of man, which is an aspiration to the highest social ideals.

In his earlier years, Karl Marx argues in the same vein that the issue of "man and nature" is a riddle of history, and the solution to it is the settlement of the antagonism between man and nature, which is, simply put, communism. He states:

> This Communism is, as perfect Naturalism, identical with Humanism, and as perfect Humanism identical with Naturalism; it is the real solution of the antagonism between man and nature, between man and man; the genuine solution of the conflict between existence and essence, between objectivisation and self-affirmation, between freedom and necessity, between the individual and the species. It is history's solved riddle and is conscious of being the solution. (Marx 77)

Despite the huge differences between the East and the West, and between modern times and ancient times, there exist extensive consensuses on "man and nature" among different peoples, who tend to consider this issue as a primary question, and even an ultimate one, though many believe that there is no such thing as the "ultimate question."

In today's reflections on modernity, the foremost crisis of modern society can be ascribable to human beings' insularity over "Nature and man." In ecological terms, the biggest mistake that man has ever committed is nothing but their hostility and atrocity against Nature.

In the works of Western scholars, "industrialization," "modern society," and "capitalism" are often juxtaposed in the same domain of knowledge. Since the commencement of modernization three centuries ago, scrutinies of and reflections upon modernity as well as criticisms of capitalism have never ceased. The most in-depth reflections and criticisms oftentimes come from within the institutions and the zeitgeist of society, covering the various aspects of a particular era or a particular social system, including politico-economic institutions, cultural mind-sets, philosophical concepts, lifestyles, and aesthetics. Reflections upon modern society boil down to a simple observation: the reversed relationship between Nature and man in this era has damaged Nature as man's home, and, even worse, distorted man's own inner nature.

Max Horkheimer (1895–1973), one of the founders of the Frankfort School, argues that the Enlightenment Movement went to extremes because, in the process of modernization, human beings' control over nature has evolved into the control of man by man. He believes that man's instrumental control over nature inevitably leads to relationships between people (Martin 294). In the industrial age, the power that controls or dominates Nature also governs people, since "the Enlightenment is identical with capitalist ideology," or in a sense, instrumental rationalism has become a capitalist ideology. Nature, mankind, and the state have become a machine serving a particular purpose. In a society like this, even literature and arts have become part of the assembly line of the cultural industry, and consequently, uniqueness, subtle sentiments, and free spirit will be rendered useless and discarded unless they are launched into the market as fashionable eye-catching commodities. The deteriorating relationship between Nature and man has corrupted the spiritual domain of mankind as well.

Martin Heidegger (1898–1976) reflects upon the nature of technology and believes that technology is not merely a means or an instrument, but rather an expression of the "relational principle" between man and nature. The real danger of the technological era does not lie in the consequences of technologies, such as nuclear weapons, but in modern technologies' landing a sharp knife deep into the relationship between Nature and man; therefore, the subsistence of Nature and man has been impaired. Heidegger cites this example: In early days, the Indians in New Mexico refused to use the steel plow for springtime plowing and deliberately took off the iron horseshoes in order not to hurt the Earth that nurtures and nourishes all. For the aboriginal Indians, the Earth was their beloved mother. In modern industrial society, however, the typical springtime farming may involve fertilizers, and herbicides and pesticides, and a roaring 100-horsepower tractor equipped with six-row two-way sharp plows in order to force the earth to yield as much as possible. As a result, the earth, once a beloved mother, has become a lamb of Passover, and human beings, once independent, have been reduced to something attached to the huge industrial machine. "Due to the will of technology, all things have inevitably become resources involved in production, before and after. The earth and the environment have been reduced to raw materials, and man to human resources, all for a set purpose" (Seubold 35). As technology reigns, the spiritual life and emotions of society are dramatically simplified; therefore, an era of plenty is also an era of deprivation. Heidegger's elaboration on "man and nature" is full of ecological concerns, informing the humanist turn in ecological studies and heralding the advent of the ecological era.

The post-1960s era witnessed flourishing ecological movements first in the West. The issue of "man and nature," or, to put it differently, the nature issue of humanity, has become the focus of public concern. Committed to ecological movements for decades, Serge Moscovici, the author of the "nature trilogy," namely *Essai sur l'histoire humaine de la nature, Society against nature: the emergence of human societies (La société contre nature)*, and *Hommes domestiques et hommes sauvages*, emphasizes in his 2002 book *De la Nature* that "the Nature issue" will be "an issue of the twenty-first century" that "in contemporary times, the Nature issue has become the very center of politics and social life" and that "Nature has become a major consideration for decision-making" (235–237). He declares, quoting Goethe, that "Nature is not a problem, but the problem" (9). He states, quoting Hans-Georg Gadamer, that Nature has become a focal point and the awareness of this situation "may be the first ray of hope at a time of world crisis" (211). He admonishes that people today should have alternative humanism, or natural humanism, because only naturalism can withstand the test of time. In a declining industrial society, Moscovici sympathizes with Max Weber, who, in turn, believes that there is eternal enchantment between man and nature, and, therefore, returning itself has a poetic touch since it awakens human beings from the deepest of their hearts, communities at the grassroots level, and life from the deepest of Nature. The enchantment is a promise of recovery which people are anticipating. Today, no responsible scholar can afford to bypass the issue of "Nature and man" in his or her discussions about politics and human destiny.

However, opinions on man and nature are divided. Barring those die-hard entrepreneurs, technocrats, and politicians, who have vested interests, serious scholars may also have different answers.

Anthony Giddens (1938–), a British politico-economist, once a counselor for Tony Blair, is a cautious historical materialist, an advocate of social progress, a restrained optimist, and a social activist with a reserved attitude toward ecological movements. In his 1998 book *Beyond Left and Right: The Future of Radical Politics*, one chapter is devoted to "Modernity under a negative sign: ecological issue and life politics." He poses an old question: How are we going to live? To answer this question, he argued modern people have to become aware that "the advance of science and technology, coupled to economic growth mechanisms, force us to confront moral problems which were once hidden in the naturalness of nature and tradition" ("BLR" 206). That is to say, the orientation and quality of human existence are to be determined by the handling of "man and nature" as the real meta-question. As a practical political theorist, Giddens hopes that human beings continue their already established modernization path and come out of plights through economic progress, global cooperation, and human beings' increased control over nature. Giddens does not agree on such voices as "returning to nature" and "reverting to tradition" as promoted by ecologists. He concludes that "we can't return to nature or to tradition." Judging from the trends of social progress, Giddens's observation seems not difficult to authenticate. But, one may still wonder whether the future life will be better than the life of today or even of the past. Can we contemplate or simply imagine things we may be unable to accomplish? Are men of letters privileged to indulge themselves in whims and illusions if practical sociologists and political scientists are not?

Giddens may not know that "Nature" and "tradition" are already part of Tao's life, as can be seen from the Tao's lines: "I retire and regain Nature, as I gave up my official life in a cage" and "Looking at white clouds from afar, I had so many longings for the past." Unlike modern scholars' reflections on modernity, examination of the natural law, ecological turn, and gaming between conservatism and liberalism, Tao's answer to the issue of "Nature and man" as the meta-question may sound too simple, but is more relevant to the origin of the issue. Besides, his answer seems to be more inclusive and zestful because of his poetic dimensions and contemplations.

In the spiritual–cultural history of humanity, Tao is not an isolated case. A prophet and explorer of the issue of "Nature and man" in ancient China, Tao should be resurrected in today's world and urge us to look far back eastward for inspirations, especially in an era bristling with so many Nature-related problems.

1.2 Tao: The Avatar of Nature

Born into a declined official clan in the later fourth century, Tao was well versed in poetry and classics, even in his childhood. He held in his prime years a lowly position akin to that of secretary, staff officer, or Head of Prefecture in the army and the government. Because of his "natural inclinations," his reluctance for his body and soul to be "enslaved by those worldly pursuits," and his distain to "bend for the meager government salary" at the sacrifice of personal dignity, Tao resigned from the post of Head of Pengze Prefecture at the age of 40. Then, he returned to his ancestral farming land and lived a self-sufficient life of Arcadian simplicity. Subsequently, he composed more than one hundred poems and essays about the *tianyuan* life, which must have livened up his toil and moil in the field.

Unlike many of his counterparts in Chinese literary history, Tao, as a poet par excellence, did not experience many arresting vicissitudes, nationally or personally, and his works do not come in multitude, but he is always admired for his personality. His works have found widespread acceptance; therefore, he has won himself an unrivalled literary status. The secret, if any, lies in his spirit, if not spirituality.

Regarding his spirit, there have been various observations among which the most common are leisure and freedom from earthly desires, abandon and aloofness, integrity, natural dispositions, accessibility, fundamental simplicity, inner tranquility and contentment, optimism and the "let-things-slide" attitude, honorable poverty (dignity in poverty), unintendedness, resignation to Nature's transformation, etc. "Nature" or naturalness may be closer to the truth about Tao, though other observations may also suffice. Nature in this case refers to the entity bearing the same name in the traditional Chinese cultural and philosophical context, meaning "the so-of-itself," pointing to the primordial unwrought state of the elements, akin to "Heaven" and "Tao" (the Way), or the sanctified Nature as Jin Yuelin highlights in his 1943 lecture in the USA as mentioned earlier.

In a word, the core of Tao's spirit is none other than Nature or naturalness. Unsurprisingly, most previous remarks on Tao's traits of character are centered on three aspects, namely the true and the natural, abandon and aloofness, and unintendedness and resignation to Nature's transformation. Ultimately, all these are derived from and made possible by "Nature" and the naturalistic spirit of traditional Chinese culture featuring the oneness of Nature out there and the inner nature of man.

1.2.1 Following the True and the Natural

Among the early individuals who use "the True" for Tao's personality and poetic quality were Xiao Tong (501–531), Su Dongpo (1037–1101), and Liang Qichao (1873–1929). In his *Biography of Tao Yuanming*, Xiao praises Tao as a man "savoring on isolation and the True"; in his "Postscript to Li Jianfu's Poetry," Su remarks, "Tao served the court as he pleased and did not hold his pursuit in abhorrence; he retired to his retreat as he pleased and did not consider his resignation as a gesture of dignity or virtue; he would beg at the doorway for food as hunger urged, or invite friends over for a meal of chicken and millets if his own minimalist creature comforts could be satisfied." People of ancient times and today pay homage to him because of his following of the "True and the natural."

In the same vein, Liang, in *The Art and Personality of Tao Yuanming*, says "Only man of Tao's like can be called a 'true man' and the art that he represents can be called true art" (16).

As a matter of fact, the True, as a philosophical concept, is a *Daoist* term which is vividly expounded in *Zhuangzi.*

In response to the inquiries about the natural and the artificial, the Sea God replies:

> That the oxen and the horses have four feet is what I mean by their inborn nature; that the horses are bridled and the oxen are led by the nose is what I mean by their enforced behavior. Therefore, it is said, "Don't destroy the inborn nature with the enforced behaviors; don't destroy the destined fate with affectations; don't damage the fame in pursuit of gains." To abide by these principles is what I mean by restoring the inborn nature. (Zhuangzi: 273)

Different from "truthfulness" in modern Chinese, the True, as opposed to the artificial, refers to the natural, unwrought state of "the so-of-itself." Holding fast to the precepts as quoted is considered as a way of returning to the natural.

On another occasion in *Zhuangzi*, the True is explored in an equally enlightening fashion:

> Confucius let out a sign and said, "Would you please tell me what natural disposition is?"
>
> The fisherman replied, "The natural disposition means the apex of absolute sincerity. Without absolute sincerity, you will never be able to elicit the emotions and passions of others, which one will never be able to touch the hearts of others. So, a forced [moan] only appears to be sorrowful; a forced rage only appears to be severe and forced warmth only appears to be friendly. A true sorrow is the silent grief without wailing; a true rage is the severity without showing the fit of anger; and true warmth is the friendliness without showing the smile. The natural disposition resides within the expression is shown without. That is why [the] natural disposition is given priority. … Following the rituals is the conduct of the worldly man; purity and innocence [are] a natural disposition, and accordingly not changeable. A man of wisdom always follows his natural instinct and is free of worldly restraints. On the contrary, a man of stupidity goes against his natural instinct and worries about worldly affairs." (Zhuangzi 557–559)

Literally, "the True" means "sincerity," which, in turn, is not to be understood simply as "true, unforced feelings and emotions" in modern psychological terms, but rather in *Daoist* terms. Lao Tzu says the following:

For the Way is a thing impalpable,
Incommensurable.
Incommensurable, impalpable.
Yet latent in it are forms;
Impalpable, incommensurable
Yet within it are entities.
Shadowy it is and dim;
Yet within it there is a force,
A force that thought rarefied
Is none the less efficacious. (45)

According to Zhu Qianzhi (1899–1972) and Chen Guying (1935–), "force" here refers to "the highest form of *qi*" (literally air) or "the tiniest primordial matter." Both Wing-tsit Chan (1901–1994) and Lin Yutang (1895–1976) translate it into "life-force," which is another form or name of *Tao*, still belonging to the category of "Heaven" or "Nature." If something that is called the True dwells within, it will naturally overflow. Even the voiceless and the expressionless are touching because they are rooted in the True, radiating from within. Psychological sincerity implied by the True is still based on Nature or the natural; therefore, there is compatibility between sincerity and the True.

In pre-Qin *Daoism*, Nature and the human mind are one, and "sincerity" is also some sort of "Heavenhood" or Nature, being in itself in the universe, or *Tao* in essence. We may find that the understandings of "the True" as "the natural," and "following the True" as "conforming to Nature," are closer to *Daoist* philosophy.

Tao makes several references to the True in his anthology. They are as follows:

In ages immemorial to man
Once lived our very earliest clan.
They lived a free and supple life,
Quite abstained from worldly strife. (Tao 27)

One sip after another drowns sorrows,
But heavens never leave me any time!
Conformity to nature is sublime. (Tao 49)

Simple longings in my heart,
With the court I would part.
Why not just let things slide,
And retreat, plow and hide.

Return to the field I may.
For ranks or fame I won't stay.
In a thatched hut I nurture the True,
And in doing so I may call it virtue. (Tao 401)

Therein lies the true meaning of life,
Veritable but ineffable. (Drinking V)

> The sagely kings are long, long gone.
> Few are returning to the True. (Drinking XX)

> As the True wanes, the artificial waxes.
> (About Scholars not Born into Due Times)

In the poetic lines cited above, the True and its combinations unexceptionally refer to the so-of-itself or inborn nature; "following or nurturing the True" means the maintenance of the so-of-itself and the individual's conformation to it; "returning to the True" means resistance to all earthly temptations and reverting to fundamental simplicity. As these lines suggest, Tao's mind is set on going back to Nature to live a natural, free life by following the True, or in other words, he intends to become a "true man," the ideal personality in the *Daoist* tradition.

In fact, in the *Daoist* masterpiece *Zhuangzi*, the concept of "true man" is expounded vividly:

> What is the true man? The true man in ancient times did not oppress the minority; he did not take pride in his accomplishments; he did not make plans. He did regret it when he committed an error; he was not conceited when he succeeded. He would not fear when he climbed a height; he would not get wet when he jumped into the water; he would not feel hot when he jumped into the fire. All these could only be achieved by someone whose knowledge had approached [D]ao. …

> The true man in ancient times knew neither the joy of life nor the sorrow of death. He was not elated when he was born; he was not reluctant when he died. Casually he went to another world; casually he came back to this world again. He did not forget the origin of his life; he did not explore the final destiny of his life. He was pleased to accept whatever came to his life; he gave no thought to life and death as if he had returned to nature. This is what is meant by not impairing [D]ao with the mind and not assisting the heaven human efforts. This is what the true man was like. Such a man had an empty mind, a calm countenance, and a broad forehead. He was as austere as autumn and as warm as spring. His joy and anger succeeded each other as naturally as the succession of the four seasons. He was in conformity with everything in the world, but he was fathomless to all.

> Therefore, the heaven and the man are one.… The man who thinks that the heaven and the man do not oppose each other is a true man.
> (89–93)

Such a "true man" has few earthly desires but strong fundamentally natural roots and transcendental traits; he defies life and death and identifies himself with Heaven. Is such a true man not a personality who is committed to Nature and becomes one with it? Ultimately, a true man is also a "natural man."

In *Zhuangzi* and *Huai Nantse*, such a personality is often deified. For instance, Zhuangzi writes:

> The perfect man is divine. Even if the great swamp were set on fire, he would not feel the heat; even if the rivers were frozen, he would not feel the cold; even if thunder hit the mountain and a whirlwind shook the sea, he would not feel the shock. Such a man rides the clouds and mists, mount the sun and the moon, and travel beyond the four seas. How can a man unaffected by life and death bother about benefit and harm? (35)

Such a perfect man, or true man, is essentially an immortal who transcends life and death, transforms all things, commits himself to the Great Chaos, places himself in the most Pure, and lives in accordance to the Way. In other words, such an immortal is someone who can rid himself of earthly desires, preconceptions, and self-complacency and therefore becomes one with the elements. According to the long-established *Daoist* idea that "Heaven and the Earth and all creations and I are one," "I" in this case is identical with the sun and the moon, the river, wind and rain, lightning, blizzards, and flames, and therefore, nothing could affect or hurt him. In a word, an immortal is somebody who is completely one with Nature.

The True and the true man, as important subtle concepts in the philosophy of Lao Tzu and Zhuangzi, epitomize the highest idealized personality in ancient China. Tao becomes an example of such a personality in the real world. However, his example is hard to follow since ordinary people usually go along a vastly different course. Consequently, there arises the conflict between man and Heaven, man's declaration of war on Nature, man's taking, hoarding and possessing, and, eventually, man's enslavery by their own worldly possessions. This is yet another case of "As the True wanes, the artificial waxes." It is certain that this will be felt more keenly as time progresses.

1.2.2 Abandon and Aloofness

In addition to "the True," abandon and aloofness are also commonly used in earlier remarks on Tao's personality and poetic gradation.

Etymologically, "abandon" in Chinese is close to vastness and profundity, and as "Wide and deep as the Valley" in Lao Tzu's term, meaning that Dao, as the Void, can contain and maintain all things. It is reminiscent of some derivations such as "do as one pleases," "abandon to fame and fortune," "tolerance and forgiveness," "the virgin land," "the vast wilderness," etc. More precisely, it implies "exile," "abandon to one's dispositions," and "self-banishment from institutions."

"Aloofness," derived from "the Void" and "the Empty" as in *Lao Tzu* ("What is most full seems empty: yet its use will never fail"), means distance, calmness, quietude, graceful effortlessness, or a pure heart.

Abandon and aloofness, perceptible but unspeakable, can be tentatively defined this way: With a pure serene heart, the individual listens to his innermost voice, follows his natural inclinations, and, therefore, lives freely and fully in the vast universe, defying earthly desires and possessions and holding fast to his independent personality and lofty aspirations. This finds expression in Tao's *A Biography of Five-Willow Gentleman* and "Homeward ho!"

In "A Biography of the Five-Willow Gentleman," Tao writes the following:

> Nobody knows his native place, his surname, or his styled name. As there are five willows growing around his house, he styles himself [as] "Five Willow Gentleman". He is reticent

and keeps himself away from wealth and distinction. He is fond of reading but reads without seeking a thorough understanding. Whenever he apprehends something between the lines, he will be so happy as to forget his dinners. He is addicted to drinking but often lacks [a supply of] wine because he is poor. Knowing about this, his kith and kin often treat him to some wine. He will drink to the last drop in their homes and get drunk at last. He will withdraw when gets drunk, never to regret having to take leave. He is content with living in his unfurnished house which cannot shelter him from wind and sun, to be dressed in ragged clothes and to see the empty baskets and gourds. He often amuses himself by writing something to express his aspirations. He has forgotten about his personal gains and losses, ready to spend his whole life in this manner. (269)

In "Homeward ho!", Tao's personally is manifested in a more straightforward way. Tao confesses:

No more of my grudges!
As there is no much time left for my earthly journey,
Why not follow my bent in living out my life?
Why all the fidgeting to seek after the unknown?
Wealth and distinction are not my aims;
Immortal lands are not my aspirations.
I only expect a fair day for me to wander alone,
Sometimes leaving my staff aside to wee and hoe.
I may ascend the eastern heights to sing a song.
Or sit a clear stream to write a poem.
I shall follow the natural cause and end my life in time;
With Heaven's decree in mind, what else am I do doubt? (Tao 247)

In the above-cited paragraphs, the key words are "oblivion" and "going alone." Oblivion, in this case, means giving up fame and fortune that those worldly people would try to woo. It emancipates the soul and elevates it to a realm of infinite freedom, independent of material things, of human agency, and of the world. A man who knows when and how to "become oblivious" requires little of the world and usually takes the road less travelled; hence, he has to "go forth alone." In his anthology, Tao treasures and chants "solitude," "isolation," "forlornness," and he likens himself to "a lone pine" or "a lone cloud" in his poetic lines: "All things have something to cling to, yet the lone cloud has nothing to lean on"; "A lone bird is on an aimless flight, even when the day approaches night … All at once it sees a tall pine-tree, and flies back from afar in utmost glee. The chilly winds have rippled the leaves of trees, save his pine-tree that grows at fullest ease."; "Why did Zhongwei live in seclusion? Because the world shared not his aspiration."; "The pure hold fast to their purity and integrity, resisting all temptations."; "Woods are not as peculiar as a lone tree." In the Chinese tradition, "going alone" (meaning independent personality) and "discovering the One" (meaning discovering *Dao*, or the true meaning of life in this case) are considered as the highest spiritual pursuits of individuals. For Tao Yuanming, "becoming oblivious," "going alone," and "discovering *Dao*" were the natural outcomes of his return to a life of abandon and aloofness in the country.

Tao's value orientation and attitude toward life epitomize the spirit of Lao Tzu and Zhuangzi. Lao Tzu says, "The Way is like an empty vessel that yet may be drawn from without ever needing to be filled" (9) and "What is most full seems empty: yet its use will never fail." (97) It means that only the pure and the empty, which is akin to *Dao*, can embrace the infinite. Zhuangzi believes that only the gentleman preserving a heart of *Dao* may become a true man, saying:

> He will let the gold lie buried in the mountains and let the pearls lie hidden in the abyss. He will not crave for property and wealth and will not strive for fame and position. He will not rejoice over longevity and will not grieve over premature death. He will not feel proud of being a high official and will not feel ashamed for being poor. He will not usurp the profit of the world as his own possession and will not regard his throne as his own distinction. He understands that distinction leads to showing off while everything in the world will return to the same root and even life and death are different phases of existence. (Zhuangzi 175)

As a result, such a man can obtain "a serene and easy heart" even in the hustle and bustle of the world and therefore becomes one with Nature, if only he is willing to appease his desires and becomes oblivious of the self. The state described here is not only the highest pursuit of such a gentleman, or a True man, but also an aesthetic realm. In fact, the much sought-after ethos of the Wei and Jin Dynasties emerged because of the interconnectedness of life's highest pursuits and the aesthetic realm in question.

Abandon and aloofness mean parting with earthly fame and fortune, following one's true dispositions, and coming closer to essentials of life. Ironically, worldly people want to hoard and keep gold as their personal belongings, or collect pearls to adorn themselves, but a true man of abandon and aloofness is willing to return the gold and pearls to where they originally belong and then go back to his rustic life. Such thought was rarely articulated in the West until reflections on modernity and ecological movements began. R.M. Rilke (1875–1926), a Bohemian–Austrian poet, is a good case in point. He writes in one of his poems that "Does the ore feel trapped in coins and gears? In the petty life imposed upon it, does it feel homesick for earth?" (179) For Rilke, the mountain is the real home of the ore, but the ore is dug out, melted, extracted, and then forged into swords, equipment, coins, and crowns to be calculated, possessed, manipulated, and looted, thus losing its true color and beauty. In his book *Earth in the Balance*, Al Gore (1948–), a renowned American environmentalist and social activist, mentions the going-home of bread. He laments that children today only know the bread as it is placed on the supermarket shelf, purchased, and served at the dinner table, but they are not aware that the field is the real home of bread, once as wheats. He states:

> Because we feel closer to the supermarket than to the wheat field, we pay far more attention to the bright colors of the plastic in which the bread is wrapped than we do to the strip mining of the topsoil. … Thus, as we focus our attention more and more and more on using technological processes to meet our needs, we numb the ability to feel our connections to the natural world. (207)

It seems that men of "abandon and aloofness" of antiquity gained a vastness and ease of their hearts and souls because they chose to return "things" to where they originally belonged; however, "materialists" today end up being enslaved, emotionally or otherwise, by "materials" as they try to transform Nature into "things." Human beings have yet to find a way to measure whether the gains outweigh the loss or the other way around.

In his earlier work *The Evolution of Aesthetics*, Li Zehou (1930–) expounds illuminatingly "the ethos of the Wei and Jin," of which Tao Yuanming is a part, and highly praises those scholars' artistic activities and achievements. He ascribes the aesthetic peak of the Chinese medieval times simply to "the awakening of man," whose core, in turn, is the awakening of "reason," characterized by "man's rediscovery, rethinking and pursuit of their life, significance and destiny" (275–276). As a matter of fact, Li Zehou tries to interpret the Wei and Jin scholars' inner cultivation from the perspective of the Enlightenment, though not without a touch of historical materialism and class struggle. He points out that the atrocities inflicted upon the scholars by the rulers eventually forced the former to retire to mountains and the woods, symbolizing "a political retreat from society to Nature." Li's argument holds water in that Nature is soothing and healing, especially for those tormented and tortured souls, as A. Schopenhauer (1788–1860) stated, "If one, who is tormented by desire or poverty and preoccupations, immerses himself in nature, he will suddenly regain a strength and courage; as a result, the waves of desire, the urges of aspirations and fears, and the desire-inflicted sufferings will be immediately soothed miraculously" (123). For "the Seven Sages of the Bamboo Grove"[1] and the Scholars of the Zhengshi Period (240–265), Nature also had a soothing effect. However, this does not apply to the Wei and Jin poets, in general, and Tao in particular, because they chose to espouse the Way of Lao Tzu and Zhuangzi, showing their real longing for a natural existence. "The ethos of the Wei and Jin" originated from "man's awakening," which means the rediscovery of Nature. At that time, being a true, supreme or holy man was identical with a man going back to Nature, or, more precisely, the latter is the prerequisite of the former. Man seems to be oblivious of the fact that the relationship between Nature and man precedes and outweighs the relationship between man and politics. Political pressure and risks did force poets to escape, but the poets' retreat out of choice reveals their conscious pursuit of the highest state of being. Besides, no trace of political persecution against Tao has been discovered in all the biographies about him. His resignation and his life in seclusion have prove that, for Tao, there is a vast, enchanting, lovable, and natural realm poles apart from the world of fame and fortune. This realm is Tao's real spiritual home, where complete abandon and aloofness are possible.

---

[1]They were a group of Chinese scholars, writers, and musicians of the third century. They gathered in a bamboo grove where they enjoyed and praised in their works of various kinds, the simple, enchanting, and rustic life. They stressed the enjoyment of alcoholic beverages, personal freedom, spontaneity, and a celebration of nature. Their life is usually considered as an antithesis of that of the corrupt court.

1.2.3 Unintendedness and Resignation to Nature's Transformation

Tao's most naturalistic thought is his unintendedness and resignation to Nature's or Heaven's transformation, transformation being the key. In his anthology, Tao mentions "transformation" (*hua,* rendered slightly differently in English for semantic consistency) many times. Some of the lines are cited as follows:

Plunge yourself in Nature's course with cheers
And you will not have any joys or fears.
When your life has reached its destined date,
There is no need complaining of your fate. (Tao 83)

My flesh has long been worn out;
A settled mind has nothing to care about! (Tao 49)

Indifferent to indigence or wealth,
No longer do I mind my wretched health. (Tao 137)

As the homebound boat goes out of sight,
I calm down, mixed with Nature in its flight. (Tao 173)

My flesh transforms with Nature's powerful force;
My spirit rests at ease in rich resource. (Tao 65)

Since the course of Nature runs its round,
By hardships are human beings bound. (Tao 67)

I shall follow the natural cause and end my life in time;
With Heaven's decree in mind, what else am I to doubt? (Tao 247)

Now that I am to die a natural death at my old age, what is there for me to linger on? (277)

According to *Ciyuan* (*The Chinese Dictionary of Etymology*), *hua* (化), or transformation, has three basic meanings: (1) change; (2) generation or formation; and (3) death or disappearance. What is special about *hua* is that it contains the transformation between life and death as two states of opposing nature. Such word formation is rare in the Chinese tradition. However, *Ciyuan*'s interpretation is not completely satisfactory because it fails to trace the origin of the word. Equally, unsatisfactory is the rendition of *Origin of Chinese Characters* by Xu Shen (58–147). According to him, *hua* means humanization or indoctrination. However, given the radicals 匕 (dagger) and 人 (man), Xu's explanation seems to be further away from the origin. If we trace further back, we have to resort to the Oracles, or inscriptions on bones and tortoise shells of ancient China. The Oracle *hua* is as follows: 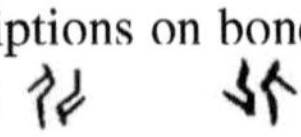

It is shaped like two men, back to back, one standing upright and the other doing a handstand. It can also be deciphered as a single man somersaulting, signifying "rotation," "reincarnation," and "ceaseless circulation," close to "returning" (Lao Tzu 59), "going back" (33), "all prevailing, unfailing" (53), and "reversal being the motion of Dao" (87). Lao Tzu articulated, "Truly Being and Not-being grow out of one another; difficult and easy complete one another. Long and short test one another; high and low determine one another. Pitch and mode give harmony to one another. Front and back give sequence to one another." (5) In fact, the positive–negative, up–down, front–back, mutually destructive, and mutually generative ever-circulating unity is the marrow of the philosophy of Lao Tzu and Zhuangzi. If *hua* is interpreted as two human figures, one *yin* (negative) and the other *yang* (positive), or as just one figure somersaulting and rotating, then it is akin to Lao Tzu's idea that "The Way out into the lights often looks dark; the Way that goes ahead often looks as if it went back." (89) "These ten thousand creatures cannot turn their backs to the shade without having the sun on their bellies, and it is on this blending of the breaths that their harmony (balance) depends." "Way is the by-name that we give it. Where I forced to say to what class of things it belongs, I would call it Great (ta). Now *ta* also means passing on, and passing on means going Far Away; and going Far Away means returning." My assumption is that the character *hua* preceded Lao Tzu, which means that the Chinese, with a long history of agriculture, noticed the ever-running life circle long time ago. Such wisdom finds philosophical expression in Lao Tzu. Chen Guying (1935–) views this as Lao Tzu's cosmology: All things, each after its kind, are full of *qi*, or life-force, flourishing and circulating. The motion of things in Nature follows the rule of negative–positive unity and eternal circulating. "Indicative of a circle, eternal circulation means ceaseless generation and growth" (Chen GY 11). Informed by the vital force of all things, Lao Tzu discovered the law of circulation ("I have beheld the ten thousand things, wither they go back.") and believes that they will eventually return to their roots ("See, all things howsoever they flourish return to the root from which they grew."). Ultimately, the roots are none other than the original primordial natural state. If the word *hua* keeps running in a circle, it will present some sort of "orderly chaos" or "organized chaos" and that may have evolved into the *Tai Chi* diagram as we know today.

Similarly, Tang Junyi (1909–1978) makes an observation that Chinese culture is characterized by "circulation and spirituality" (Mou 14). The Daoist circular viewpoint of Nature, allegedly derived from the oracle character *hua* as mentioned above, reminds us of modern theories about the "biosphere" and the "ecosystem."

As is known, E. Suess (1831–1914), an Austrian geologist, posed the concept of "biosphere" in 1875. It was the first scientific attempt to view all things as an organic whole, evolving in a sphere that is almost identical to a circle. "Biosphere" has emerged to become a basic term in eco-studies. In the same vein, A.G. Tansley (1871–1955), a British botanist, put forward the concept of "ecosystem" in 1935. He believes that all living things in nature exchange information and energy with the environment and among themselves in an interdependent, interactive, and automatic feedback process. This idea is also close to the ever-running "transformation" process, which is the most natural, primordial state of being. The Chinese has a long-established idea of "all natural beings as the so-of-itself being in a perfect state of being."

Suess and Tansley noticed air, rocks, water, woods, pastures, grazing herds, insects, etc., in this "system" or "sphere," but, unfortunately, ignored human beings. For them, human beings seem to be spectators outside the sphere rather than players in it. Thinkers of ancient China, however, always kept in mind that human beings are also a part of "Nature's transformation" and of the ever-evolving biosphere. Only when human beings are aware of their being part of the transformation will they be firmly rooted, inspired, and granted freedom and intrepidity. This wisdom is carried further forward in *Zhuangzi*:

> The great earth endows me with a physical form to dwell myself in, makes me toil to sustain my life, gives me ease to idle away my old age, and offers me a resting place when I die. Therefore, to live is something good and to die is also something good. (Zhuangzi 96)

> Therefore, the sage dwells in [D]ao where nothing will be lost and thus lives with it forever. Yong and old, alive and dead, the sage always serves as a model for all. How much more important is [D]ao, on which everything in the world relies and every change depends! (Zhuangzi 95)

> Therefore, it is said, "he who understands heavenly joy follows nature when he is alive, changes with everything in the world when he is dead, shares the virtue of yin when he is still and shares the movement of yang when he is active." It follows that he who understands heavenly joy will not be complained by heaven, will not be blamed by men, will not be entangled in worldly affairs, and will not be reproached by ghosts or spirits. Therefore, it is said, "he moves like a heavenly body and he is still like the earth itself. When his mind is settled down, Heaven and the earth get into normal order. His physical form does not suffer from illness and his spirit will not get weary. When his mind is settled down, everything in the world will submit to him." (Zhuangzi 203–205)

All these boil down to "the oneness of Nature and man" and "the oneness of life and death."

Tao's lifetime of life is refined in his poetry into an ecologically loaded philosophical–aesthetic experience as shown by his commitment to Nature and his resignation to Nature's transformation. For Tao, transformation means his resignation and commitment to Nature; it also means his eternal serenity.

Modern ecological studies hold that a corporal being is always part of the "circulating" of the ecosystem since living organisms always interact with the environment in substance and energy. The elements of the cosmos are also those of our bodies. The liquid circulation within the human body contains the sea, lakes, and rivers as wind and storm contain human respiration. Our corporal bodies return to Nature as they come from it. Similar thought can also be found in *Zhuangzi: Perfect Happiness*:

> But if we trace her (my wife's) beginning, she did not have life before she was born. Neither did she have life, nor had she physical form at all. Neither did she have physical form, nor had she vital energy at all. Amid what was opaque and obscure, transformation took place and she obtained her vital energy. Another transformation took place with her vital energy and she obtained her physical form. Yet another transformation took place with her physical form and she obtained life. Now that one transformation has taken place and she has returned to death, this is like the succession of spring, summer, autumn and winter. (Zhuangzi 289)

For the individual life, the body just returns to Nature for another journey of circulation after life ceases to be. Vaguely, among the ancient Chinese, Tao, for one, already discovered that law. In his poetry, Tao writes, "As death replaces every life for sure, an early death can't be called premature." "As I knew my fate was predestined, I never worried about my gains and losses." This reveals Tao's vision and graceful resignation to Nature's transformation.

1.3 Dwelling Poetically in Nature

From Tao's choices and pursuits, we may identify such a path: from honorable poverty to fundamental simplicity, then to the True, and then to the Natural. Unlike the progress-oriented path of modern society, it is a path not to be taken by the "reasonable" majority today. Tao did pay a price for his course, the price being, among others, his lost power and distinction, his toil and moil in farming, his forlorn life, and his occasional hunger and cold. But what he gained were his relief from, and transcendence over, life and death, a serene mind, an amity with the elements, an ease and freedom, and the love and worship from later generations though he himself never craves so. It is safe to say that Tao's lifetime pursuit was to "dwell poetically." Therefore, his life may be better sketched as follows:

Returning to the country lifekeeping integrity in poverty-keeping to fundamental simplicity_following the True and the Natural_dwelling poetically.

According to Zhao Yifan (1950–), dwelling, or *wohnen* in German, means the harmonious coexistence of man and Nature (158). Indisputably, a poetic dwelling is also a life of bliss conducive to Nature and harmless to none. For many, as they often confess, a reading of Tao's poetry may enable them to forget their nobility or humbleness, or life and death, and, at the same time, help them to "exorcise their earthly ambitions, complacency, and self-pity; therefore, greedy men may become disinterested, and cowards independent" (Xiao Preface).[2] Such a confession is proof to the soothing and transforming the power of Tao's poetry, for which the feats of heroes such as the emperors of the Qin, Han, Tang, and Song or Genghis Khan (the Supreme Conquer) are no substitutes.

Speaking of poetic dwelling, I must refer to Heidgger's ontological poetics. In his reading of Hölderlin (1770–1843), he discovered something so close to Tao Yuanming's spirituality. He cites a few lines from Hölderlin's 108-line-long *Homecoming*:

> Gateway prompts me to go on home instead,
> Where the busy highways are familiar to me,
> To visit the countryside and beautiful valleys
> Of the Neckar, and the forests, where godlike green
> Oak and beech trees and silent birches gather, and
> A friendly spot in the mountains still holds me captive. (Heidegger 13)

For Heidegger, the "homeland" refers to such a space that allows the individual a place where he can have a sense of "being at home" and, therefore, exists therein, such a space being a bestowment of the earth intact. The sky and the earth

[2]Xiao, Tong. Preface to the *Anthology of Tao Yuanming*. Rare Ancient Edition.

of the homeland are the "angels" for all creations. Heidegger argues that the most beautiful and fundamental about the homeland is that it, more than anything else, is closer and more faithful to the origin. What is homecoming, then?

For Heidegger, homecoming is coming close to the origin. One may have a feeling of *déjà vu* and then come to a sudden realization that it is exactly what Tao means by "returning to the rustic life of fundamental simplicity."

Heidegger thinks that the homeland in Hölderlin's poetry is "Engel des Hausses" (angel of homes) and "Engel des Jahres" (angel of time), which are similar to "Heaven and the Earth" or "transformation" in classical Chinese philosophy, or "Nature's God" in Jin Yuelin's term. It is in this sense that Heidegger declares that the homeland is the root and origin of the soul. The soul must dwell herein as trees take roots in the earth. Heidegger believes that the poet's poetic dwelling precedes that of men in general, and, therefore, the poetic creative soul is as always "being at home" and close to the origin. In Tao Yuanming, the same idea is expressed in a much simpler and more accessible way in his poetic segments like "indifferent to fame and fortune," "following natural inclinations," "oblivion of gains and losses," "keeping to the rustic life," "becoming a poetic hermit-farmer," and "resignation to Nature's transformation." If there has to be a more concrete example to prove how the poet's soul dwells poetically at home, we may try to feel for Tao in his following self-narration:

> I have been fond of liberal arts since my childhood. On occasions, I would take pleasure in having a little leisure. Whether I apprehend something between the lines, I would be so glad as to forget about my dinners. I would be overjoyed when I see the overlapping shadows of the trees or hear the bird-songs in different seasons. I used to say that when I lie in leisure beside the northern window in May or June, I would feel like a man living in [the] ancient times. (271–273)

What Tao describes is his "being-at-home" state, home being not only about the thatched hut, but also the earth and the sky of his homeland which has tremendous natural charm. As one can imagine, the poet, "dwelling" by the northern window, is free from earthly distractions, enjoying the bestowments of Nature and becoming one, body, heart, and soul, with Nature. I feel attempted to ask, what else is out there worthy of man's aspiration and pursuit, if it is not poetic dwelling?

In a sense, poetic dwelling is a passage to the realm in between Heaven and the earth, a space, actual or invoked, where man becomes one with Nature, a realization of spiritual value in aesthetic experience, and a most valuable being in life and in the universe alike. However, such being has long been overwhelmed by something practical, instrumental, or material. Due to the declining of Nature and the drying of poetic inspiration, the younger generations in today's world do not have much experience of poetic dwelling. Therefore, I hereby cite the experiences of two "enlightened" persons, namely Li Yu and John Muir, in hopes of striking a chord with the younger souls.

Li Yu (1610–1680), a Chinese poet of the late Ming and early Qing dynasties, was forced to bid adieu to his "normal life" at a time of turmoil, only to find himself merrily immersed in Nature. His experience is as follows:

> When the Ming court fell and the Qing was about to emerge, I, determined to be far away from any distinction or official position, lived in a mountain to avoid the outside upheaval and became proud of my idleness. I did not visit friends in summer, nor would they come over. Since nobody else was around, I took off my shirt, pants and shoes, let alone my scholar's scarf. I would either stay naked under exuberant lotus leaves, and therefore my wife had difficulty spotting me, or, alternatively, lie on my side under the towering pine tree without even noticing a passing ape or a crane. I would wash my inkpad in the downpouring spring, and try making tea with melted snow. If I felt like eating gourds, they were simply out there; if I wanted to have fruits, they were hanging low on the branches. This could be called a life of heavenly novelty and supreme pleasure. Later on, I moved to the city and had to handle many social engagements. I felt myself to be burdened by vanity, though my life there was least about any money-making industry. In retrospect, I spent only three short years living a life akin to that of immortals. (335)

Presumably, this romantically inclined soul might not know that his experience is almost identical with the life of Adam and Eve in the Garden of Eden.

Similarly, John Muir (1838–1914), a vanguard of the eco-movement of the world, equally attached to Nature, writes in his field report about his lingering in Nature. His report overflows with poetic enchantment:

> We are now in the mountains, and they are now in us, making every nerve quiet, filling every pore and cell of us. Our flesh-and-bone tabernacle seems transparent as glass to the beauty around us, as if truly an inseparable part of it, thrilling with the air and trees, streams and rocks, in the waves of the sun – a part of all nature, neither old nor young, sick nor well, but immortal. (78–79)

In his *Our National Park* (1906), Muir tries to capture his experience in Nature:

> Nature's peace will flow into you as sunshine flows into trees. The winds will blow their own freshness into you, and the storms their energy, while cares will drop off like autumn leaves. As age comes on, one source of enjoyment after another is closed, but Nature's sources never fail … The miracle occurs … One day is as a thousand years, a thousand years as one day, and while yet in the flesh you enjoy immortality. (56)

Muir's experience of the transformation of things and the self is almost identical with "the oneness of Heaven and man" in *Daoist* philosophy. One tends to feel inspired by the striking likenesses in poetic experience and values between people of the West and of the East, and between modern Western people and ancient oriental people.

1.4 The Nature and Naturalness of Tao's Poetry

Although "style is the man" as conventional wisdom is almost completely deconstructed by scientific literary criticism, it still remains a fundamental though simple preposition for honest literary theories. It is already an established observation that Tao's poetry is natural, and its naturalness is derived from his natural dispositions.

Firstly, Tao's subject matter is predominantly Nature, especially the *tianyuan* presences, among which the most common ones are the woods, mountains, the sun, the moon, winds and frosts, rains and dews, cypresses and pines, elms and willows, mulberries and sesames, chrysanthemums and peaches and orchids, birds, crops, chickens and dogs, the wooden fenced door, the thatched hut, the well and the kitchen, the neighbors, humble food, and spring wine with sediment of course. In cases where some things do not seem that natural, the artificial is kept to the minimum, as we may agree that a thatched hut is no rival for a well-appointed mansion in terms of artificiality.

Secondly, Tao's poetic style is naturalness. Criticism of various dynasties on Tao's poetic style includes "fundamental simplicity," "authenticity," "purity," and "graceful ease," to name but a few. All of these are implied by "naturalness," meaning returning to the Simple and the True, or the primordial state of Nature. Interestingly, when critiquing, earlier critics tend to liken Tao's style to natural presences out there. For instance, Yang Wanli (1127–1206) remarks, "Tao's poetry is like spring orchids, autumn chrysanthemums, winds through pine trees, and water in the valley." (*The Integrity Study Collection*, Volume 80) Zhu Dianpei (1418–1491) states, "Tao's works are like egrets over a clear stream and elks in the woods." (*The Pine-Stone Study Remarks*) Chen Zuoming (1665?–?) says, "Tao's poetry is like a gorge in autumn, with clouds afloat, leaves fallen, water clean, the sun setting hill-ward, and the vault vast and silky. Seeing the gorge's simple plain color at the first glance, the viewer may think that he has seen the whole picture without seeing its changing ethereal states that are not readily accessible." (*Caishutang Selected Poems*, Volume 13). Similar metaphors include "exuberant riverside grass after the rain"; "purple red clouds changing freely"; and "eternal ease of the Milky Way, and ever purity of lotuses in a clean pond."

A further question may be asked here: Why could Tao manage to produce so "natural" poetry? And what is the significance of his "Nature only" poetry for him and later generations?

Bai Juyi (772–846), for one, asks, "Why were Tao's artistic conceptions so profoundly subtle?" Standing on the Xunyang Tower, Bai came to a sudden realization: Tao's inspirations come from the grand chilly river, the towering mountains, the moon over the pools, the morning mist over the peaks, etc. (*Poem composed on the Xunyang Tower*). What Bai means is that Tao draws his vitality and inspiration from his enchanting and enlightening homeland Nature. In eco-critical terms, the ecological position of the subject-of-a-life exerts heavy influence on the color and the nature of the subject.

With a profound understanding of the Wei and Jin literary ethos and scholars' mind-set, Luo Zongqiang (1932–) expounds on the naturalness of Tao as a man and of his poetry. He argues that, apparently, the Wei and Jin scholars were all close to Nature and tended to rest their feelings and emotions on landscape, but, in essence, they belong to different categories depending on their intentions and eventually realized aesthetic experiences.

The first category included nobles as represented by "the Jingu celebrities." They tried to seek higher spiritual pleasure that their luxurious creature comforts could not offer, and infused their landscape complex into their sensual indulgences in order to have "both sensuality and nobility." The Jingu Dale banquet-tour of Shi

Chong (249–300) and his *coterie* are a good case in point. The Jingu Dale used to be a resort on the outskirts of Luoyang city. With his overwhelming political power and wealth, Shi Chong had a garden built in the dale. In the huge garden, there were clean streams, ranges of hills, water-pillowing houses, woods, and scattered mansions, pavilions, and platforms. There were also fruit trees, vegetables, medicinal herbs, game animals and birds, and even waterpowered pestle and deliberately shabby huts and caves that could add a rustic touch. The owner and his guests would stand atop the hill and look down afar, or sit by the brook angling. Their life indoors was made all the more zestful by bells and music. Their luxurious endearment to Nature had to be supported by status and wealth. Therefore, for them, Nature was reduced to a consumption commodity or a fresh-flavored dessert after they were satiated with the main courses.

The second category was the "Lanting Literati," whose peer leaders were the clansmen of the Wang Xizhi (303–361) and the Xie An (320–385) clans. They were gentries with literary talents and refined aesthetic tastes. After they moved to the south from the fallen north, they wanted to have a peaceful spiritual life and refined deportment. As a result, they retired to Nature as their spiritual underpinning and soothed their souls therein; hence, there came the much talked-about "gathering at Lanting (Orchid Pavilion)." Wang Xizhi has a vivid account in his preface of "AT THE ORCHID PAVILION."

> In the background lie high peaks and deep forests, while a clear, gurgling brook catches the light to the right and to the left. We then arrange ourselves, sitting on the bank, drinking in succession from the goblet as it floats down the stream. No music is provided, but with drinking and with songs, our hearts are gay and at ease. It is a clear spring day with a mild, caressing breeze. The vast universe, throbbing with life, lies spread before us, entertaining the eye and pleasing the spirit and all the senses. It is perfect.

Luo Zongqiang rates this account very high, saying, "A few sketches capture the real aura of the landscape there … They shifted from perceiving Nature to contemplating life" ("MM" 273). Obviously, these literati like Wang Xizhi had a higher aesthetic experience than the Jingu celebrities. However, Luo Zongqiang makes reserved remarks on the former, because they were just appreciators of pleasing Nature. The beauty of Nature was of importance in their life, but there was a gap between them and Nature after all. In other words, the relationship between them and Nature was characterized by a subject–object division, rather than by real oneness.

Luo pays the highest tribute to the impoverished Tao instead. He believes that unlike the first two types of scholars, Tao becomes one with Nature. Luo says, "In Chinese cultural history, Tao was the first to become one with Nature. He was part of Nature rather than a viewer, not to mention a possessor. He was not merely close to Nature, but in it." ("MM" 274) Luo continues, "Nature in Tao's poetry is rural Nature, nurturing simple dwellers in it. Or it is an organic community of Nature and Man … Tao's heart beats to the rhythm of Nature, because, ultimately, Nature and Tao's body and spirit are just one" ("MM" 275). Informed by the philosophy of Lao Tzu and Zhuangzi, Luo goes a step further and identifies the spirit of Tao's poetry, saying, "The oneness of the self and things, and of the heart and Nature, is the highest Daoist ideal, and of metaphysics as well. But, ever since the rise of metaphysics, no one except Tao has ever realized that state … because Tao resigns himself to Nature's transformation in earnest" ("MM" 275).

From a psychoanalytical point of view, Tao might have inherited more of the collective unconsciousness of the Chinese nation, which is vaguely called "a misty emanation" or "a misty mass of chaos" in traditional Chinese critical discourse. The collective unconsciousness is naturally evoked and represented in Tao's poetry. This may help to explain most of the remarks of earlier scholars on Tao's poetry. Natural, well-conceived, and close to "the roots," Tao's poetry is destined to have a universal appeal.

In Jung's psychology, collective unconsciousness is often depicted as an old man with his race's millions of years of wisdom. There is also such an old man in Tao's unconsciousness, as He Yisun (lived around 1637) believes, "Tao is an enlightened one who has acquired the Way, and his poetry has achieved eternity though he did not crave so" (252). Regarding how one can find *Dao*, Zhuangzi says the following:

> I acquired it from the son of Literacy, who acquired it from the grandson of Recitation, who acquired it from Insight, who acquired it from Comprehension, who acquired it from Diligence, who acquired it from Chant, who acquired it from Profundity, who acquired it from Emptiness, who acquired it from Creation. (99)

The acquisition of *Dao*, it seems, involves the investment of the self and a shift from the visible to the invisible and the other way around. Chanting, which is in the middle of the process, is akin to ballads or poetry. Above it is consciousness, which is more visible, and beneath it is an abyss of unconsciousness, whose origin is an unknowable, irreducible Beginning, or creation, or "a mass of chaos" (Chen GY 219). In his *A Collection of Notes on Zhuangzi*, Guo Qingfan (1844–1896) interprets Beginning as "the root of *Dao*," "a mystic realm," or "the Doorway whence issued all Secret Essences." This explains why Tao's poetry, though not plentiful, has eternal appeal and vitality. The secret lies in its rootedness in the origin of *Dao*.

1.5 Tao's Greatness Out of Nature-Centrism

Any Chinese literary history textbook may contain such a statement that "Tao Yuanming was a great poet." A question now ensues: In what way, to what extent, and by what means? If I may use "impression," or "viva voce (word of mouth)," a somewhat loose concept, though, to gauge the individual's historical status, Tao can well take rank with the First Emperor of the Qin (Qin Shihuang).

As a status indicator, "going down in history" has long been considered by the Chinese as supreme glory. "History" can mean voluminous annals or word of mouth. In this sense, Tao and Qin Shihuang, seemingly poles apart, can be bracketed together.

Accredited by annals of all dynasties "the emperor of emperors," Qin Shihuang inaugurated highly centralized dictatorship to be followed by all the subsequent imperial dynasties. By contrast, Tao, crowned "the poet of poets," has heavily influenced poets of later generations. Besides, Qin Shihuang is considered as the starter of "the imperial seal" to be inherited; Tao has enkindled a literary torch to be passed on. Historians and literary critics aside, people at the grassroots level surely know of Tao if they know of Qin Shihuang as they definitely know about "the Peach-blossom Springs" if they know about Qin Shihuang's "burning books and burying scholars alive." Interestingly, Liang Qichao believes that no praise for Tao would suffice and, therefore, produces not without funny abruptness, a very plain remark: "This man's status is way too awesome!" (3) "Status" in this case does not mean "official rank," but rather "historical status" and "personality," since Tao, even as Head of the tiny Pengze Prefecture, cannot match Qin Shihuang, almost identical with Heaven, in terms of social status or by otherworldly standards.

However, when it comes to history-making, Qin Shihuang and Tao differ vastly in their personal effort as well as in social costs and the historical price.

With the legacy from his predecessors, Qin Shihuang spent his whole life fighting, conquering, and slaughtering, and his title "the First Emperor" was baptized in bubbling blood. According to *Shiji* (*Historical Records*) of Si Maqian (145–86 B.C.), the Kingdom of Qin decapitated 1.81 million people in the 22 of its 93 wars to annex the other 6 kingdoms. The total number of people slaughtered in the 93 wars was about 4 million, almost one tenth of the Chinese population at that time. In fact, the powerful Qin became a highly efficient killing machine, and for this reason, visitors, when appreciating the terra cotta warriors and chariots (excavated in Xi'an) of Qin Shihuang, should not forget his atrocities. After having annexed the six kingdoms, which was an exploit in itself, Qin Shihuang, in addition to his "upholding justice," "enforcement of law," and "standardization of measurements and the writing system," also enacted severe punishments, including cutting the body into pieces, cutting off the nose or feet, decapitation, castration, execution by five horses (the body and the four limbs are torn apart by five horses), cutting open the abdomen, and body-boiling in a huge wok, so as to establish his absolute authority. He decreed his exclusive use of the words "*zhen*" (meaning "I"; used by the emperor to address himself.) and "*xi*" (the emperor's seal). To show its nobility and eternal glory, he toured afar and had his deeds inscribed on stone tablets and had luxurious palaces built, which involved hundreds of thousands of prisoner-laborers and exhausted most of the rare old trees in Sichuan, Hunan, and Hubei provinces. In fact, the entire Chinese nation paid a huge price for the First Emperor.

In stark contrast, Tao has made history more "easily," at least apparently—simply by drinking, reading, plowing, and chatting, composing poems about the southern mountain and chrysanthemums, fumbling on his possibly unstringed zither, etc. All was plain and simple. As a poet, he only wrote a little more than one hundred pieces of poetry and prose.

The result of the comparison, however, may be surprising. Qin Shihuang, like a towering castle or palace on one end of the balance, and Tao, like gentle breezes or white clouds on the other end, should weigh almost the same, though intuition would favor the former. As always, opinions on Qin Shihuang are divided and many believe that his demerits outweigh his merits. On the contrary, Tao enjoys generous praises, except for a brief spell of a curse as cast on almost all ancient saints during the Cultural Revolution period. In writing history, people have got used to highlighting something tangible such as politics, economy, and military achievement and, in the meanwhile, ignoring something less tangible such as gentle breezes and white clouds in man's spirituality.

But, what on earth are gentle breezes and white clouds?

Simply put, gentle breezes and white clouds are "the climate," the issue that frustrated the more than one hundred heads of states at the Copenhagen climate conference on December 7, 2009. Essentially, they symbolize a nation's unique spirit and aura, or its inner nature. Seemingly light, feeble and empty, gentle breezes, and white clouds are no less important than steel-and-cement buildings.

On the balance of humanity, Qin Shihuang and Tao represent different views of the world and life. It is necessary to go back to Jin Yuelin's observation if we want to talk about world views more wisely and more effectively.

During his stay at the Oriental Institute of the University of Chicago in 1949, Jin wrote a book in English, titled *Tao, Nature and Man*. This book is an attempt to present with discourse familiar to the Western academic community the spirit of traditional Chinese culture. In the chapter "Nature and man," Jin proposes the "simple vista," the "heroic vista," and the "sagely vista" as mentioned earlier.

For Jin, a simple vista is characterized by the dichotomy or bifurcation of reality as well as that of self. Therefore, it is closer to man's childhood which is full of playful enjoyment and free from ego-centric predicaments. At the same time, it is independent, self-sufficient, and desires little; therefore, a man having such a vista can have inner peace and composure.

A heroic vista points to an internal–external and self-things division. Man alienates himself from Nature, lords over others, and considers the external world as his antithesis or something to be conquered. Such a man is ambitious, highly capable, calculative, strong-willed, and decisive, and often cruel. He fears no challenge or sacrifice, and his mind is set on achieving success in his career. He is ready to promote progress and change the world. Jin believes that without heroes, human civilizations would stagnate, and this is proof enough to heroes' indispensability. But, the heroic vista is only an aspect of human nature. For the world, heroes alone do not suffice.

With regard to the sagely vista, Jin writes:

> The sagely vista is somewhat like the naïve except that its apparent naiveté is arrived at through advanced meditation and contemplation. The behavior of a man with a sagely vista may appear naïve as a man with a naïve vista, but the discipline behind it is based

> upon meditation that transcends the human function with the result that one is not merely free from ego-centricity but also of anthropocentricity. ("TNM" 189)
>
> …
>
> The only way to do is to know one's destiny, to be at peace with one's station in life in a much more comprehensive than the merely social and political sense. Obviously, some desires will have to be satisfied, but if in satisfying or trying to satisfy them, one is at peace with one's destiny, one is reconciled with one's limitations. Whatever destiny or station that is given, it is for him a charge to keep, a function to fulfill in the democracy of mutually dependent co-existents. He admits neither self-satisfaction nor self-depreciation. He does not struggle against what is within his function, nor does he crave for what is beyond it. Since he is a man, he has to realize the essence of his being, and whatever is given to him in addition as a separate self, he accepts with as much good race as he accepts his humanity. What is needed is not saintliness which involves being Godly and, therefore, other than man, but sagacity which involves the transcendence of the mere nature in man as to approach that nature in man which is also t'ien. The latter is within the possible reach of any man. Once it is reached, one may be in vista different from most men and yet in every other aspect the same as any John, Dick or Harry. In him, object nature and subject nature are unified and in the unification there is harmony. ("TNM" 195)

For Jin, those having a sagely vista never abuse their rights, wealth, knowledge, or wisdom. They, superior to ordinary men, are completely immersed in ordinary men as ordinary men, just like any Tom, Dick, or Harry. They treat their destiny with calmness and composure and try to pursue inner peace, integrity, and social harmony. Their life may be despicable in materials terms, but fundamentally simple, natural, and poetically blissful.

We may realize that the sagely vista is exactly what Tao espouses. A general survey of human history may reveal that there have been numerous heroes but few saints. Heroes make or mar, as saints do or choose not to do. Most of the plights and disasters of modern society have something to do with the feats of heroes, in one way or another.

After an anatomy of the three vistas, Jin points out that ever since the era of Greek mythology and the Holy Bible, "the West seems to be dominated by the heroic vista," whose basic principles are "ego-centricity" and "anthropocentricity." Driven by such a vista, the West has made remarkable achievements in politics, economy, and military. But, such a vista can be "destructive" because, in ecological terms, the predominance of man over nature is maximized, and the natural world is disappearing. Consequently, in the West, people are literally "caged" in an "artificial environment of artifacts." As a promoter of human society and a transformer of Nature, such a heroic vista and exploit are oftentimes "appalling."

The rapid change of the world is fascinating and awe-inspiring. But people seem to be always preoccupied by the possibility of going astray, since each and every one is under pressure, because they are plagued by ever-growing desires, and therefore rendered helpless. The outcome, in the best-case scenario, may be the disappearance of poverty and suffering in economic terms. However, this does not mean that happiness is bound to come and grow as expected.

Based on his observation of the social ills, Jin arrives at a conclusion as follows:

> The fundamental problem of today may be one of social legislation or economic rearrangement so as to better the conditions of life, but that of tomorrow is bound to be one of individual salvations as to improve the quality of living. What is needed is not a class of sages, but a section of people in different functions achieving the sagely vista. Socially as well as individually the trouble is not with our stars but with ourselves, and to prevent the social organism from being dominated by the heroic vista which is going to sweep the world, it is necessary to permeate it with sagely vista. (197)

1.6 Nature: The Intersection of Two Philosophies

Definitions of "nature" are confusing, especially in modern times. Joachim Radkau (1943–), a German historian on the world environment, observes that subject to the influence of philosophers, people tend to take nature simply as a concept rather than as intuition or as the experience of the fundamentals of life. He says, "As biological organisms, human beings and the other organisms follow the same natural law: they will die from lack of water, starve to death if there were no fauna and flora, shrink if there were no sunshine, and go extinct eventually if there were no intercourse. Nature is not merely a subject for discussion, but ultimately the origin that defines the animal nature of human beings" (22). Radkau's opinion is insightful in that, for him, Nature is more of an attitude toward the world than of an abstract concept.

R.G. Collingwood (1889–1943) points out that in remote ancient times, especially at the source of human civilizations, people of the East and of the West tend to have similar attitudes toward Nature. For instance, Thales (624–574 B.C.) of ancient Greece compares Nature to a "cow." His conception of Nature is "enormously remote from the Renaissance conception of the natural world as a cosmic machine made by a divine engineer in order to serve his purposes." He regarded it as a cosmic animal whose movements, therefore, served purposes of its own. The animal lived in the medium out of which it was made, as a cow lives in a meadow (Collingwood 32).

Similarly, Lao Tzu likens Nature to a "Mysterious Female," an infinite mother of all things, which reminds us of Gaia. He believes that "The Doorway of the Mysterious Female is the base from which Heaven and Earth sprang." (13) In fact, both oriental natural philosophy and its Western counterpart tend to treat Nature as an organic, chaotic, and generative entity that is enticing and enchanting and awe-inspiring, full of vitality, and having will and emotions of its own. Wang Guowei (1877–1927) argues that "matter" in ancient Chinese philosophy is biological and has an intrinsic life. He exemplifies this with "物" (*wu*, or things) as in "物理" (*wuli*, or law of things, or physics). The radical form of "物" being "牛" (*niu*, or cow), physics in Chinese is none other than a "study of the cow," reminiscent of the "Profound Female" in *Lao Tzu*. "Law of things" as such is also physiology and, further, philosophy of life. Ancient Chinese philosophy is a natural philosophy that is centered on life activities. Unsurprisingly, Aristotle's zoology contains knowledge of physics, which is proof enough to the blurred boundary between physiology and physics in the earlier natural philosophy of the West.

In classical Chinese thinking, the oneness of Nature and man is indisputable. For ancient Chinese, the human body is part of Nature and humans "must respond to Heaven and the Earth, and *yin* and *yang*"; there is no definite distinction between human nature and beastliness, or, more precisely, "there may be beastliness in humans and humaneness in beasts"; even politics and Nature work on each other, as can be seen from such quotes: "Observe the Heaven, and you will know the current politics. Observe human activities, and you will humanize the world."; or "The saint's governance for peace is reliant upon the harmony between Heaven and Earth." Besides, man's literary and artistic activities are also one with Nature: "Literature is crucial in nurturing virtues because it is one with Nature … Thoughts and feelings evoke words, and words lead to literature. This is the Way of Nature." In contrast, it seems that there were much fewer records in the pre-Socratic West on the relationship between Nature and human society. However, these limited and incomplete records point to an organic holistic view of the universe. For instance, scholars at that time had a similar quest, that is, to discover the origin of the universe. Thales (624–547 B.C.) believes that the originating principle of Nature is water; Anaximenes (585–525 B.C.) thinks that it is "air" and that the concentration and dilution of air have generated the cosmos; Heraclitus (540–470 B.C.) asserts that the origin is "fire," and the whole world is an "eternal fire," ever changing though. Zeno of Elea (500 B.C.–?) goes one step further and argues for the connection between Nature and virtues. For him, to live according to Nature is to live on virtues, which is reminiscent of the spirit of *Lao Tzu.*

In the post-Socratic era when scholars became "sophists," the principal concern of philosophy shifted from the quest of the origin of the cosmos to that of morals, ethics, and knowledge. In Plato, the ideal world and the physical Nature are separate, opposing worlds, and such dichotomies as substance and spirit, body and soul, essence and phenomena, content and form, individuality, and universality took form. Aristotle, then, laid the foundation for Western rationalism and scientism by metaphysics, formal logic, and scientific classification. Only in modern times, especially after Francis Bacon, Descartes, and Newton, was Nature completely concretized and materialized; therefore, it has become an "object world" as opposed to the human world. Nature is widely considered as resources that provide for humans' well-being, or as a machine following its own law. Humans have become the very center of the world, and they dominate Nature in many, if not all, ways.

Since the Enlightenment Movement and the industrial revolution, the West has witnessed rapid modernization thanks to its rationalism, scientism, and market operations on a quantitative basis. Its flourishing economy, improved social welfare, scientific progress, affluence, and power are considered as a model of happy life to be admired and emulated. Consequently, as Western materialization spreads rapidly, a modern view of Nature has found widespread acceptance ever since.

In China, the classical view of Nature still remains, though there may be differences between different schools. Traditional Chinese culture never diverges from the underpinning conception of the oneness of Nature and man; however, it inherits and repeats itself, thus ending up having deep roots but not much foliage, metaphorically. Unlike their Western counterparts, the Chinese people clings to the identity between Nature and human society and this has led to a stagnating and humiliated China, once a perfect example of "ignorance and backwardness."

In Chinese tradition, there did appear a materialistic, rationalized, and practical view of Nature, which is an effort to distinguish man from Nature, as seen in Xun Zi (298–238 B.C.). Xun Zi declares that "Order or disorder has nothing to do with Heaven," so "Superior men should be clear about the Heaven-man division." He argues, "Is it not much better to heap up wealth and use it advantageously than to exalt Heaven and Earth and think about it? If we neglect what man can do and think about Heaven, we fail to understand the nature of things." (Fung "SHCH" 234) Unlike the Confucian Heaven that has will and inner value, or the Daoist Heaven in the ontological sense, Heaven for Xun Zi is a materialized entity that is akin to today's "Nature." In his *A General History of Chinese Thought*, Hou Wailu (1903–1987) also points out that "The Confucian view of Heaven changed in Xun Zi from a willed Heaven to a Natural Heaven" (I 531). Xun Zi elevates "human agency" to a very high status in terms of the cosmological hierarchy and believes that it "can demonstrate its power," "rule things," and "offset or utilize the mandate of Heaven" (536). Xun Zi maintains that Man should confront Nature, try to control it, and promote social progress by machinery and skills. Unsurprisingly, he is rated as the zenith of pre-Qin philosophy by historical materialists. A comparison of Xun Zi and Francis Bacon, two millennia the former's junior, would reveal a number of similarities between them, especially in their attitudes toward Nature. Wing-tsit Chan (1901–1994) asks why has China failed to develop its natural science as the seventeenth-century Britain did, even though Xun Zi already established the conception of Nature in explicit terms? This question is thought-provoking, indeed. If the Chinese nation had abandoned Confucianism and Daoism for the naturalistic course of Xun Zi, would China have been the first to enter into the industrial age? If industrialization had commenced two thousand years ago, what would have become of the earth? Is it a blessing or a curse that China has been "creeping" for so long in an agricultural society because of its discarding of the naturalistic course of Xun zi and its embrace of Confucianism and Daoism? There is no ready answer to these questions.

Xun Zi per se is proof enough to the existence of the materialized, instrumental, and utilitarian view of Nature in China. The course of Xun Zi was followed by many scholars of later generations, such as Zhang Heng (78–139), Wang Chong (27–97), Fan Zhen (450–515), Liu Zongyuan (773–819), Liu Yuxi (772–842), Dai Zhen (1724–1777), and Zhang Taiyan (1869–1936), to name but a few, but it never reigned in any dynasty.

Only after the Western powers forced open the door of China at the end of the nineteenth century did some Chinese "radicals" realize the power of science and technology. As a result, they began to part with their ancestor's spiritual legacy and began to turn to the West for "truth," and to advocate "human rights," "rationalism," "science," endeavor, and man's triumph over Nature. The statement of Chen

Duxiu (1879–1946) was the most typical during the May 4 Movement time. Chen says:

Along with the establishment of the Darwinist evolution theory featuring natural selection, survival of the fittest, and God's absence in man's making, human beings, enlightened now, begin to believe in their triumph over nature by their intelligence, to establish principles and maxims through scholarly exploration, to realize that they are responsible for their own blessing or curse and course of action, and to discard their excessive religiosity or blind resignation to God. Thus, the European society has marched by leaps and bounds … So, the Chinese people should get started to make a difference by exalting science and human rights, if they want to rid themselves of their shallowness and ignorance (Chen DX 8–11). Ever since the May 4th Movement, it has been China's consistent state policy to follow the modernization pattern of the west, upgrade its science and technology, and accelerate the exploitation of natural resources in order to become an "industrial" country, despite power shifts. Some were skeptical about such modernization drive. For instance, Gu Hongming (1857–1928), Du Yaquan (1873–1933), Xiong Shili (1885–1968) and Liang Shuming (1893–1988), among others, rejected both the idea of a thorough westernization and a complete departure from tradition. They profess that oriental culture should be rebuilt on the basis of the reconciliation between the East and the West and a compromise between modernity and tradition so as to promote material and spiritual civilization as well as the harmony between Nature and man. Ironically, they were labeled as conservatives and counterrevolutionaries, thus slipping into an abyss of helplessness and distress. Worse still, China just let slip an opportunity for real cultural reciprocity between China and the West, especially in terms of the notion of Nature. Deplorably, this mistake turned out to be a hidden trouble for China's subsequent social progress.

Unexpectedly, when the Chinese follow in the West's footsteps in its swaggering modernization drive, Westerners have already begun to reflect upon their own course in a serious fashion. The early nineteenth century witnessed the rise of criticism of instrumental rationalism and a rethinking of modernity and science and technology that ushered in the industrial age. Reflections on the Western view of Nature have been part of the ethos. Such reflections have made a great number of prominent thinkers: Rousseau (1712–1788), Thoreau (1817–1862), Nietzsche (1844–1900), Dilthey (1833–1911), Simmel (1858–1918), Whitehead (1861–1947), Scheler (1874–1928), Schweitzer (1875–1965), Spengler (1880–1936), Jaspers (1883–1969), Leopold (1887–1948), Toynbee (1889–1975), Horkheimer (1895–1973), Heidegger (1889–1976), Marcuse (1898–1979), Bertalanffy (1901–1972), Cobb (1925–), Rolston (1933–), and Griffin (1939–), to name but a few. Even some scientists have joined the reflection movement: Einstein (1879–1955), Bohr (1885–1962), Heisenberg (1901–1976), Monod (1910–1976), Prigogine (1917–2003), etc. Informed by this reflective type of thinkers, Western ecological movements have emerged and culminated, marked by the publication of *Silent Spring* by Carson (1907–1964) and the proposal of the Gaia hypothesis (also known as the Gaia theory or the Gaia principle) by Lovelock (1919–) and Margulis (1938–2011). Thus, "Nature" has regained its high status it deserves in the West.

In this wave of reflections, some Western scholars, free from that typical Western pride and contempt, have begun to study Chinese culture and appreciate ancient Chinese philosophers in earnest, so as to draw upon the experience of the ancient Chinese regarding the man–nature relationship. Iliya Prigogine (1917–2003) emphasizes that "Chinese civilization represents a profound understanding of the relationship among man, society and nature," and "Chinese philosophy is a constant source of inspiration for those scientists and philosophers who want to broaden the scope and significance of Western science" (1–2). Joseph Needham (1900–1995) declares that Lao Tzu is the man who knows Nature best and Daoism in Chinese culture is still active and vital. After having read "Let Be and Let Alone," a chapter in *Zhuangzi*, Needham states with finality that "Bearing in mind what mankind knows today about soil conservation and nature protection, and all the experience we have gained as to the proper relations between pure and applied science, this passage of *Chuang Tzu* (*Zhuangzi*) seems as profound and prophetic as any he ever wrote" (199). Similarly, after quoting the opening sentences of "The Movements of Heaven" in *Zhuangzi*, Prigogine professes with equal excitement that "We believe that we are moving toward a new synthesis and a new naturalism. We may eventually synthesize Western tradition (with its emphasis on experiment and quantitative description) and Chinese tradition (with its spontaneous, self-organizing worldview" (57–58).

In stark contrast, the entire twentieth-century scholarship in China, with the exception of a few non-mainstream scholars, had dramatically decreased enthusiasm about nature contemplation. Reflections in China on the modern industrial society also came much later than in the West.

Over the past two decades, in response to the rising eco-movement and the needs of the times, some scholars have begun to explore natural philosophy, and, in particular, the idea of the unity of the natural world and the human world contained in traditional Chinese culture. They believe that "the oneness of Nature and man" is the greatest of ideas that traditional Chinese culture has to offer and that such an idea serves as a valuable academic resource that will contribute to the handling of the world's ecological crisis. In utter solemnity and sincerity, Zeng Fanren (1941–) states:

> Built into the consciousness of the Chinese nation, the conception of the oneness of Nature and man, a hallmark of Chinese culture, is the very marrow of traditional Chinese thought. The ecological wisdom contained therein is of extreme significance; therefore, we have to do justice to it. Recognizing and exalting the value of this idea means our identifying ourselves with Chinese culture in a global context. If we abandon those ideas as essential as the oneness of Nature and man, the Chinese cultural identity will be further blurred, and the Chinese would end up being homeless spiritually and emotionally. (443)

As "reversal as the movement of the Way" in Daoist philosophy and "the law of the negation of the negation" have proved time and again, things, when at an advanced stage, may return to their most original state. In fact, when critiquing "the modern society," those postmodern thinkers tend to sympathize with their pre-modern counterparts, and therefore, they become time-transcending academic soulmates. "Returning to the pre-Socratic era," a voice from some postmodern philosophers epitomizes such an appeal.

Undoubtedly, China has richer and more complete pre-modern cultural and spiritual resources than ancient Greece in the pre-Socratic era. Besides, Chinese cultural continuity is well maintained over the past two millennia. It comes as no surprise that in order to better reflect on modernity, some Western philosophers begin to turn to Chinese classics for inspiration. It is worth noting that after the emergence of existential phenomenology, "Metaphorically, classical Chinese scholarship is no longer a dwarf or an awkward singer out of tune with the orchestra; it becomes more lively, enticing, smart, and pleasant, and, therefore, people feel that it is vital and has much to offer" (Zhang XL 8). Interestingly, Schweitzer cites the story about "the old man holding an urn" in *Zhuangzi* to convey his idea that modern society does harm to people's morale and spirit. His conclusion is as follows:

> The dangers that were suspected by the gardener in the fifth century B.C. are active among us in full force. Purely mechanical labor has become the lot of members among us today … We are all more or less in danger of becoming human beings instead of personalities. Therefore, many kinds of material and spiritual injury to human existence form the dark side of the achievements of discovery and invention. ("CE" 268–269)

The most essential of questions that the academia, East or West, faces is to rethink and readjust the relationship between man and nature in order to alleviate, if not solve, the world's eco-crisis and to promote peace and harmony. As we enter into an ecological era, we cannot afford to refuse those academic and spiritual resources regarding fundamentally simple phenomenology, primordial holism and generation, harmonious Nature aesthetics, and spontaneous eco-philosophy in the Chinese tradition. Unlike the past century when the Chinese scholarship tended to "look westward," the new century will see a more proactive role of the Chinese scholarship thanks to its inborn appeal and the needs of the times, heralding the beginning of its independence, if not global influence.

As a Western scholar states, "The 21st Century will see, for the first time, the true globalization of philosophy… Asian philosophy will play a major role in philosophy in the twenty-first century." (Priest 85–99)

Starting from *The Book of Change*, "Nature" has remained the point of departure and the zenith of classical Chinese thought as represented by ideas such as "the production and reproduction is what is called 'change'," "the misty mass of chaos gave birth to all things, each after its kind," "*Dao* generates all things," "*Dao* follows the so-of-itself," "man is one with Heaven," and "love is to be extended to people and all creations alike." Ubiquitous and essential, such ideas combined define classical Chinese philosophy. For Chinese thinkers, Nature is "the origin of all kinds of generations," "the time-space where cosmological transformations take place," "the creation containing reason and enthusiasm," or "the unity of the human existence and the cosmological life." All of these boil down to one observation: The Chinese tradition is rooted in an oriental type of "natural philosophy." Given the gravity of "Nature," Tao Yuanming, as a Nature poet and avatar of nature, is destined to resurrect in a new synthesis of oriental and occidental philosophies.

Chapter 2
Tao Yuanming's Natural Philosophy

Since the May Fourth Movement in 1919, scholars, armed with modern science, tend to treat Daoism and Confucianism the same way as they treat the idealism-materialism, subjectivity-objectivity, and consciousness-substance dualities. However, these dualities and the metaphysical mode of thought they epitomize do not apply to studies of classical Chinese philosophy.

Presumably, the Confucian admonishment of "Manage the nation in order and let peace prevail" was planted in the mind of Tao Yuanming during his adolescent years due to his exposure to, and immersion in, Confucian classics, since Confucianism rose to become the mainstream orthodox state ideology in the Han Dynasty and has remained so in imperial China. In the meantime, Tao Yuanming, influenced by the ethos of the Wei and Jin dynasties, absorbed the refined thought of Lao Tzu and Zhuangzi and the metaphysics and, as a result, developed his own cosmology and view of life. Considering his natural dispositions, life experiences, and aesthetic preferences, he might have immersed himself in the profundity of Daoist philosophy, which finds expression in his poetry. These observations, logically sound though, are not free from the Confucianism–Daoism dualism. They appear too superficial when applied to Tao. Interestingly, the views of Tao from earlier scholars, ambiguous though, seem to be more valid. For instance, Liu Chaozhen remarks, "Neither Confucian nor this-worldly, neither egocentric nor defiant, neither gallant nor stubborn, Tao is contented, free from the unnatural, he has got all Nature can offer."[1] Zhong Xiu states in the same vein that "(when it comes to remarks on Tao) one does not have to be obsessed with whether Tao is identical with Lao Tzu and Zhuangzi, or enlightened immortals or not. He is by far the only one who is completely unrestrained without going astray."[2] What Zhong means is that Tao is what he is: philosophical, given to Nature, sympathetic with

[1]Tao, Shu. Tao Shu's Collection of Remarks on Tao Yuanming's Works (Rare Ancient Edition).

[2]Zhong, Xiu. Tao Yuanming's Anecdotes and Poetry (the Qing Dynasty Edition).

S. Lu, *The Ecological Era and Classical Chinese Naturalism*,
China Academic Library, DOI 10.1007/978-981-10-1784-1_2

Creations, and following Nature's transformation or the "so-of-itself." Such a man, like *Dao* itself, beggars description.

Strictly, *Dao* is not to be interpreted merely as a *Daoist* concept, but as the core of natural philosophy that originated in times of antiquity. It finds expression in Lao Tzu's and Zhuangzi's cosmology, Confucian Heaven-*Dao* theory, Mohism, the School of Logicians, the Yin-Yang School, and, later on, the (Buddhist) Zen practice. Tao Yuanming does not simply belong to Daoism, or Confucianism or Buddhism; he belongs to Nature. In this sense, he and his poetry are the avatars of classical Chinese natural philosophy.

It is safe to say that natural philosophy can be a good approach to Tao as the avatar of Nature and that Tao may offer new insights into contemporary natural philosophy. However, a thorny issue has to be addressed: What is natural philosophy and what is naturalism?

Since dictionary definitions are plentifully confusing, we may as well pick the simplest one: Natural philosophy or naturalism is a philosophy that the cosmos, society, and life are observed, explained, and experienced with Nature as the yardstick and from the viewpoint of Nature. Understandings of and experiences with Nature vary from historical phase to historical phase, and, therefore, their intention is not monolithic. Pre-modern natural philosophy is best represented by "the oneness of Nature and man" in ancient China; modern natural philosophy by the dichotomy of man as the subject and nature as the object as promoted by Francis Bacon and Sir Isaac Newton. Natural philosophy here in this book is used as opposed to anthropocentrism; it holds that nature is also a subject in juxtaposition with human beings, who are just creations out of Nature, and that the cosmos is an organic community consisting Heaven, the earth, gods, and man. Natural philosophy in this sense is none other than "ecological philosophy," as Edgar Morin (1921–) states:

As regards anthropology, ecological studies restore the high status of "nature," and roots life in nature. Nature is no longer a disorderly, passive, amorphous environment, but rather a complex whole. For this complex whole, man is no longer a closed entity, but an open system residing within in it in an autonomous-dependent fashion in organizational morphological terms. (14)

Ecological studies proceeding in the humanist direction is, in essence, post-modern natural philosophy. In this context, our way of thinking and our approaches should be amended accordingly.

In modern times, the most serious harm to nature is inflicted by extreme rationalism. Does this mean that orthodox scholasticism and the way of thinking it represents should be changed? Can studies have an extra touch of intuitional experience and poetic contemplation since nature can be viewed as intuitional experience?

Bernard Stiegler (1952–), a student of Jacques Derrida, says that the most essential is also the most familiar, but, in our culture, it becomes the remotest and invisible. Truly, this has become a common problem: The simplest is the least visible. For instance, for fish in water, water is the least felt because they are in it. In a sense, Nature is to man what water is to fish. Stiegler's declaration is thought-provoking:

> Hearing becomes a question, but for natural law to be natural, it is indeed necessary that everyone hear and understand it immediately, originally, without having to use strange and studious reasonings. Natural law must be before reason itself, indeed before reasonings (those of philosophers and metaphysicians as well as those of everyman). It is reasonable to think that everyman, learned or not, can hear it. But the more learned he is, the more difficult it will be for him to hear, for his culture will obfuscate the naturalness of the law (reason, "by its successive developments", will suppress nature.) … The more natural it is, the deeper it is hidden in the "appallingly ancient": in order to remember, to recover the evidence prior to the fall, reason must be forgotten. Earlier than I think, I am because I feel, I suffer. (109–110)

From cradle to grave, we are exposed to culture and called "men of culture." However, should we not be alert to the risk of the "voice of Nature" being muted by boisterous culture? To avoid this, we may try to follow in Tao's footsteps.

2.1 Tao's Philosophizing

A philosopher poet of ancient China and today's Sinosphere, Tao Yuanming deserves a place in the pantheon of the philosophy of the world, not because his poems, limited in number though, are also philosophy at the same time.

Chen Yinque, among others, made an early attempt to evaluate Tao from the perspective of natural philosophy. He believes that Tao is a great poet and thinker because he inaugurated "neonaturalism" of the medieval times. Based on an examination of the social milieu of the Wei and Jin dynasties, Chen points out that Tao's philosophy, having a trace of "just follow the so-of-itself or the Already-there" as demonstrated by Ruan Ji and Liu Ling, two of the "Seven Sages of the Bamboo Grove," is different than Ruan's and Liu's indulgence in sensual pleasures and complete freedom, and that though Tao does not belittle Confucianism, he has no intention of wooing worldly glory because of his inclination toward naturalness and of his resignation to nature's transformation. Therefore, "Tao Yuanming is a Confucian without and Daoist within" (Chen YQ 229). Chen's observation is widely recognized by later scholars. It is safe to say that Tao's philosophy is akin to "fundamentalist" Daoism.

In his 1997 book titled *A Study of Tao Yuanming*, Yuan Xingpei (1936–) scrutinizes Tao's "natural philosophy." Yuan's philosophical approach leads to some pioneering results. It is worth pointing out that philosophers are also involved in the studies of Tao Yuanming. In his 1940s monograph dealing with Nature and life, he shows his preference for classical Chinese natural philosophy in general and the "spiritualized Nature" in particular, though he criticizes the *Daoist* concept of nature from the Hegel's point of view. For He Lin, Tao's poetry proves the poet's internalization of Nature and, therefore, the oneness of Nature and man. He writes:

> In modern times, what we mean by "going back to Nature" is return to the spiritualized Nature instead of soliciting self-denial or self-annihilation, or a blind indulgence in the natural world out there. Precisely, it means that humans internalize Nature and let it glow in their souls. The southern Mount in Tao's poetry, the Peach-blossom Springs, and the

> landscapes in paintings and poetry, in general, can be considered as that nature which glows in souls. It epitomizes the oneness of Nature and man rather than a blurring of their boundaries, or hostility between them. The oneness as such represents humans' "spiritual conquer" of Nature instead of material conquer, or, in other words, it means a nature that is elevated by the human spirit. (He Lin 122)

Zhang Shiying (1921–), a contemporary Chinese philosopher, carried further forward the philosophical approach to Tao's thought. In his *The Interaction between Heaven and Man*, which goes beyond the demarcation between philosophy and literature, Zhang, from an East–West comparative/contrastive point of view, juxtaposes Tao Yuanming and the later Martin Heidegger and finds that the two sages meet in terms of the origin of "being." His statement is as follows:

Tao's poetry has a philosophical touch, and therefore, barring *Daoism*, which is Tao's inspiration, Heidegger's philosophy seems to be the only Western philosophy that is akin to Tao's philosophy and can explain Tao's poetry. Heidegger's philosophy has a poetic inclination, and therefore, it seems that Tao's poetry is the only Chinese poetry that can explain Heidegger's philosophy (Zhang SY 375).

Heidegger's philosophy, in this case, refers to existential phenomenology, which, as a point of departure, may lead us closer to Tao's poetic philosophy.

In the 1930s, Heidegger acquainted himself with Daoism and translated in collaboration with Xiao Shiyi (1911–1986) a number of chapters of *Lao Tzu*. Heidegger holds as his antithetical motto a sentence taken from Chapter 15 of Lao Tzu, which reads "Which of you can assume such murkiness to become in the end still and clear? Which of you can make yourself insert to become in the end full of life and stir?" However, in Heidegger's works, Tao Yuanming is not mentioned because Heidegger might not have known such a Chinese poet, nor could Tao predict the existence of Heidegger, who is a millennium Tao's junior, but they meet in *Daoism* conveyed by *Lao Tzu* and thus become "soul mates." Why *Daoism*, then? In the domain of philosophy, their roads lead to the "origin." Tao returned to his country life, symbolizing his poetic return to the origin of the simple and the natural; Heidegger returned to the pre-Plato and pre-Aristotle Greek philosophy when human beings and "being" were one. This era coincides with the golden era of Chinese philosophy when Lao Tzu and Confucius established their philosophies.

However, contemporary times have witnessed a suffering nature and declining poetry, which pose a threat to human beings, physically and spiritually. This seems to be an irresistible trend. Heidegger in his twilight years intended to become a banner of philosophy that could save the world on the edge of collapse. He tries to uncover the origin of nature's decline in modern times on the one hand and to trust the saving of the world to literature in general and poetry in particular in a techno-Benthamite era. Heidegger's later return to the realm of arts shocked the academic community because he not only added zest to philosophy but also turned the course of philosophical studies. Ever since Heidegger's philosophical turn, "the concept of nature" and "poetic dwelling" have resurfaced as major philosophical issues of the times. Some even hope that Heidegger's philosophy can help us find a way out for modern people who seem to be disoriented before they are beyond salvation.

In his endeavor to explore a path of salvation, Heidegger was hugely inspired by F. Hölderlin (1770–1843), a German romanticist poet, which symbolizes the amity between philosophy and poetry in a post-Plato era. Heidegger, who is so un-Western in terms of philosophical tradition, reinterpreted Nature, poetry, and the relationship between poetry and Nature from the viewpoint of existential phenomenology. His philosophical turn has some sort of oriental aura found in philosopher poets or poet philosophers.

In his *Elucidations of Hölderlin's Poetry*, Heidegger explains that he chose Hölderlin instead of Homer, Vigil, Dante, Shakespeare, Goethe, etc. though their works could reveal the nature of poetry as well. Heidegger believes that Hölderlin's poetry is a witness to humans' attachment to the land and nature, and that Hölderlin, as a poet, gracefully shuns the situation that "[d]anger is the threat beings pose to being itself" (55), thus enabling language to lead humans back to the origin of the world and to a spiritual realm where beauty and serenity reign. Such language can only be poetic language. Heidegger emphasizes that "Poetry itself first makes language possible. Poetry is the primal language (Ursprache) of a historical people," (60) and that the being of human beings is essentially poetic. Human "say" in poetry by virtue of language without hurting nature, and this is, in Heidegger's term, "the most dangerous activity" (So vulnerable nature is!). However, for Heidegger, poetry and only poetry can lead people back to the roots of nature, where people can enter into infinite serenity. Therefore, Heidegger declares that poetry is the most innocent undertaking.

Hölderlin writes the following lines:

> Whom no master alone, whom she, wonderfully
> All-present, educates in a light embrace,
> The powerful, divinely beautiful nature. (75)

Heidegger offers a detailed interpretation: "The inner movement of these three lines strives toward the final word 'nature', and there the movement comes to its end. … Nature comes to presence in human work and in the destiny of peoples, in the stars and in the gods … in streams and in thunderstorms. 'Wonderful' is the omnipresence of nature" (75). Heidegger then argues that "nature embraces the poets. They are drawn into this embrace. This inclusion transposes the poets into the fundamental characteristic of their being. Such transposition is education. This characterizes the poets' destiny" (77). Judging from his rendition, Heidegger has broken away from quintessentially philosophical discourse and, along with Hölderlin, entered into poeticized, soothing, holy nature.

In order to correct the direction of history development in an era of mounting crisis for human existence, and to save the land and human spirit, Heidegger trusts his hopes to poetry and eventually discovers Hölderlin. Heidegger does so because Hölderlin, the poet of poets, is a true poet in Nature's soft soothing arms. In fact, philosophers' poetry and poets' philosophy or philosophizing, among others, are both profound ways to reveal the secret of human existence.

Do Heidegger's praises for Hölderlin not apply to Tao as the avatar of Nature? Admittedly, Heidegger might have no idea that in the far Orient some 1600 years ago, there lived a poet, or, a soul out of Nature and Muse.

The decline of Nature and human spirit today is even more serious than Heidegger's day. A cold comfort may be that "where there is crisis, there is salvation." In a context of growing ecological movement, the Chinese should consider it their responsibility to present to the world their Tao Yuanming, or, the world's Tao Yuanming, who may be part of the salvation of the world.

2.2 The Significance of Nature in the Sinosphere

The earth is being jeopardized by ever-increasingly frequent ecological disasters. As a result, "Nature" has become a common concern for the public and a key word for eco-critics.

The twentieth century witnessed an unprecedented trauma for Nature both in philosophical and physical senses in the East and the West. In response, there have risen some strong eco-voices such as "reverence for nature," "preservation of nature," and "returning to nature."

Nature, like "culture," is a seemingly accessible everyday expression in Chinese. However, a closer scrutiny of it, which is like a labyrinth, has to invoke different historical stages and diversified cultures. This expression seems to contain all secrets about the universe, the human world, and all forms of literature and arts. It is necessary to reveal the meaning of Nature and its usage in the Sinosphere before moving on to discuss Tao's natural philosophy.

According to most Chinese dictionaries, the interpretation of "nature" (自然, *ziran* in *pinyin*, meaning "self-so") is usually "heaven-so" (天然, *tianran* in pingyin, meaning literally "Heaven has made it so.") or "non-artificial." This explanation is far from being satisfactory, because, literally, Nature, as a noun, may have a better definition than an adjective. However, the interpretation of "self-so" as "heaven-so" is worth exploring. The key to this is how "heaven" is understood. In pre-Qin Chinese philosophy, "Heaven" (*tian*) does not mean the "sky" over the earth, but the supreme, infinite, absolute entity that has a will of its own, which generates and dominates all things out there, each after its kind. Heaven is *Dao* (the Way), or otherwise called "the *Dao* of Heaven." Since the *Dao* of heaven runs of itself, it is called "self-so." Lao Tzu's statement that "The ways of men are conditioned by those of earth. The ways of earth, by those of heaven. The ways of heaven by those of Tao, and the ways of Tao by the Self-so" (Chapter 25) does not mean that the self-so is something that Heaven follows. Being autonomous or subject to their individual laws, Heaven, Tao, and the self-so are identical. It is in this sense that "Nature" has some sort of "heaven" at its very source and functions as a substitute for Heaven or *Dao*. It has evolved into an absolute being that has infinite vitality, purposefulness, autonomy, and generating power. The wisest of creations as humans are, they have to revere and follow Nature.

The English word "nature," which is widely considered as an equivalent for Chinese *ziran*, meaning "the so-of-itself," is fundamentally different from the latter. Though the English-language "nature" does have a touch of autonomy and self-existence, it mainly refers to the self-existent physical world in time and space as opposed to the human world. Precisely, in this sense, it is only a rough equivalent to "the ten thousand things," meaning all creations, in classical Chinese. Well versed in both English and Chinese, Jin Yuelin uses "the unification of nature and man" in English to refer to "the oneness of Heaven and man" in Chinese. However, Jin also states that Heaven as in "the unification of Heaven and man" is much more meaningful than "nature" in the sense of the natural world. For Jin, nature's God in the West is closer in meaning to Nature in the Chinese sense. He believes that the difference between Nature in the Chinese sense and pure nature (the physical world) is that man belongs to the former but is alienated from the latter (Jin 151). Similarly, in his elucidations on Hölderlin's poetry, Heidegger compares "nature" with φύσις and concludes that man "transposes subsequent elements into the beginning, and replaces that which is proper to the beginning with something alien to it" (79). Therefore, Nature in classical Chinese deserves further exploration in today's eco-criticism.

It is widely acknowledged that the mainstream thoughts of classical Chinese culture are Confucianism and Daoism, whose fountainhead is generally believed to be *The Book of Change*, which, in turn, descended from the fortune-telling custom of the Shang Dynasty. As the unearthed oracle characters suggest, fortune telling at that time falls into two categories: Nature and human activities. The former includes calendar, meteorology, climate, time sequence, and direction, human activities; the latter includes farming, hunting, and wars. Confronted with the meta-question of "Nature and man," people would resort to the unity of opposites such as up and down, front and back, in and out, motion and motionlessness, blessing and curse, victory and defeat, auspiciousness and omen, and life and death to explain their confusions about life and the world. What makes fortune-telling possible is the faith in the telepathy and interaction between gods and man as well as that between nature and man.

It is a consensus of the academia that *Daoist* philosophy mainly deals with the operation of the universe or Nature, and Confucianism society or human activities.

The most influential of Chinese natural philosophical masterpieces are *Lao Tzu* and *Zhuangzi*, especially the former. The following statement best captures the role of Daoism in Chinese culture.

Without *Lao Tzu*, Chinese culture and the Chinese national traits would have become vastly different. In fact, even Confucianism, the mainstream thought of Chinese society would not have been the same, because, historically, it was influenced by Daoism as Chinese Buddhism was. Without a real understanding of the profundity of the little book (*Lao Tzu*), one is not expected to understand Chinese philosophy, religion, politics, medicine, and arts including culinary art (Chan 137).

Chan believes that among all the Chinese classics, *Lao Tzu* is the most influential though the shortest. According to statistics, there have been more than 700 hundred versions of interpretive works, including over 40 English versions. In Lao Tzu, "*ziran*" is mentioned 5 times. They are as follows:

> But from the Sage it is so hard at any price to get a single word
> That when his task is accomplished, his work done,
> Throughout the country everyone says:
> "It happened of its own accord." (Chapter 17)

> To be always talking is against nature. (Chapter 23)
> The ways of men are conditioned by those of earth.
> The ways of earth, by those of heaven.
> The ways of heaven by those of Dao, and the ways of Dao by the Self-so. (Chapter 51)

> No mandate ever went forth that accorded to Dao the right to be worshipped,
> Nor to its "power" the right to be worshipped,
> Nor to its "power" the right to receive homage.
> It was always and of itself so. (Chapter 51)

> And so teaches things untaught,
> Turning all men back to the things they have left behind,
> That the ten thousand creatures may be restored to their Self-so.
> This he does; but dare not act. (Chapter 64)

"*Ziran*" listed above unexceptionally means "the self-so," "the naturally so," or "the primordially so." In Lao Tzu's logic, only Dao is the unfathomable, unspeakable self-so which is the root of Heaven and the Earth "whence issued all the secrets." In this sense, Dao is almost interchangeable with "Nature."

Ziran is mentioned many times in *Huai'nantse*. In his *Notes* to *Lao Tzu*, Wang Bi (226–249) almost identifies *ziran* with *Dao*, saying "As Heaven evolves, Dao is identical with *ziran* …What we mean by '*ziran*' is but a representation of the unspeakable and the infinite" (*Notes to Lao Tzu*, Chapter 25); "Heaven and the earth simply let it be, and because of their non-artificiality, all things run themselves. Therefore, Lao Tzu said 'Heaven and Earth are ruthless; to them the Ten Thousand things are but as straw dogs'"(*Notes to Lao Tzu*, Chapter 5); "All things follow ziran; therefore, they are not to be forced" (*Notes to Lao Tzu*, Chapter 29); and "It is not that the actual representation of *ziran* is unseen, but that its being is not be seen. Nothing can change what it is" (*Notes to Lao Tzu*, Chapter 17).

Hou Wailu summarizes Wang Bi's natural philosophy this way: "In generative terms, nature is not physical, but absolute; it is but an expression for the infinite; therefore, it is still a master, the master of all things; it is speechless and intangible, and therefore should not be treated as a subject of cognition, but only as something to follow" (III 115). Such natural philosophy as held by Wang Bi holds that Nature is a hidden intangible being independent of any external force. It dictates all physical creations.

Such a conception of Nature as a formless infinite intangible existence independent of external forces but dominant over all other existences negates human's capability of knowing and grasping Nature, and opposes the materialization and objectivation, and then the transformation and conquering of Nature. It maintains that human beings can only follow Nature and be one with it so that they can act without going astray, do without being too artificial, which can be best described by "With non-doing as the model and speechlessness as the teaching, things, each after its kind, would have their true nature, which is identical with *Tao*" (*Notes to Lao Tzu*, Chapter 23).

In essence, Lao Tzu's natural philosophy as expounded by Wang Bi is close to Western existential phenomenology. Due to historical and ideological restraints, Hou, close to spelling out Tao's natural philosophy, fails to expand his vision and modify his logic so that he labels the subtle philosophizing found in classical Chinese philosophy as sheer "idealism" in the negative sense of the word.

Unlike Hou, Wing-Tsit Chan has a more down-to-earth and unprejudiced stance in his treatment of traditional Chinese cultural spirit. In his elaboration of Lao Tzu's natural philosophy, Chan summarizes the significance of Daoist natural philosophy in three aspects, namely the conception of Nature, that of society, and that of life.

Conception of Nature: *Dao* is the one; it is natural, eternal, spontaneous, nameless, and indescribable; it is the origin of all creations, the process of motion, and the underpinning for the operation of all things.

Conception of society: When individualized by specific things, *Tao* becomes a virtue. The individual's ideal life, the ideal social order, and the ideal form of government should be based on *Dao* and governed by it.

Conception of life: When it comes to the way of life, *Dao* means the simple, the spontaneous, the tranquil, the feeble, and, most importantly, inactivity. Inactivity does not mean "doing nothing" but "doing nothing that is against Nature." In other words, "it is a process of nature unveiling itself" (Chan 137–138).

2.3 The Modern Evolution of Daoist Natural Philosophy

The twentieth century witnessed the almost entire collapse of Daoist natural philosophy in China due to the impact of the various kinds of revolutionary campaigns and the modernization drive. Outside China, Daoist natural philosophy is only found in a few Chinese scholars overseas, whose stances and attitudes seem to change as the world's philosophical ethos change. Lamentably, Daoist philosophy has survived the hard times thanks to those scholars that are labeled as "neo-Confucianists."

In his *A General Study of Lao Tzu and Zhuangzi*, which contains eighteen essays on Lao Tzu and Zhuangzi spanning over 30 years, Qian Mu (1895–1990), a master of Confucian studies, touches upon the Daoist conception of Nature. As a self-taught, grassroots-to-high caliber scholar, he, at a time when his nation and culture were on the verge of collapse, was decidedly committed to curing the ills of the time. To this end, he resorted to Confucianism, among others, as the cure, advocating "to save the world through China; to save China through Confucianism," a motto he had espoused from his prime to twilight years. Obviously, his theory is different than the May Fourth Movement slogan "Down with Confucianism," yet they share the same purpose of re-energizing the nation. Qian Mu's attitude to Daoism is inevitably defined by the ideology of the times. In his essay titled "The Conception of Nature in *Guo Xiang's Notes to Lao Tzu,*" Qian compares Wang Bi's and Guo Xiang's notes regarding the conception of Nature and finds that the former believes that "Nature generates all" and that "Nature is but a substitute for the nameless and the infinite, the primordial and the formless"(Qian 368). For Wang Bi, *Dao*, Non-being, and Nature are identical. This idea sounds not sound to Qian, who does not believe in "something out of nothing." Guo Xiang argues that things cannot be generated out of nothingness and that things evolve because they are what they are. For Guo, Nature is virtually "the natural world" and there is no sovereign over Nature. In short, Wang Bi emphasizes "non-being" and, therefore, non-doing and Guo Xiang "being" and therefore human agency; the former leads to the Way of Heaven and the latter points to "the law of things." On this point Guo Xiang and Qian Mu converge, and the latter generously praises the former's conception of Nature, saying, "Guo Xiang's interpretation to the conception of Nature as in Daoism is the most relevant, accurate, and paramount. It is no exaggeration to say that naturalism in Daoism was established by Guo" (Qian 369). For Qian, Guo Xiang's naturalism not only developed Daoist naturalism but also overtook the latter to become the summit of classical Chinese natural philosophy. It must be pointed out that his personal preferences aside, Qian tries to interpret Daoism from the perspective of Confucianism and to let the latter governs the former. Consequently, he partly succeeds in transforming Daoist natural philosophy via Guo Xiang into part of the practical solution to the problems of the times and even led it onto a modernist utilitarian and rationalistic track. In his series of books, Qian exalts Confucius and belittles Lao Tzu because he is committed to the idea that scholarship should practically benefit the state, the citizens, and social progress in general. His pursuit as such coincides with the ethos of Enlightenment and rationalism of modern Chinese society.

Fang Dongmei (1899–1977), four years Qian's junior, also dedicated to promoting Confucianism, treats "Nature" as a proposition in a very different way.

Unlike Qian, who is more like a "pure Confucianist" who tries to transform the "impractical" part of Daoism with orthodox Confucianism in order to facilitate social progress, Fang, well immersed and versed in modern Western philosophy, and with Henri Bergson (1859–1941), Freud, Whitehead, etc. as his spiritual mentors, has a better understanding of the inadequacies of some modern Western conceptions and the accumulated ills of the Western modernization, and, therefore, has more academic though sometimes complicated dimensions. He is committed to unveiling and correcting some of the ills of modern Western society and thus became an early part of the modernity reflections in China. In order to achieve his goal, Fang, as a "neo-Confucianist," assumes a more open profile as can be seen from his proactive absorption of natural philosophy from the Daoist canon. The following quote clearly shows his methodology and goal:

> Nature, as the name of it suggests, should refer to the natural world. Ontologically, it is an absolute being and the root for all existences. Primordially, it is identical with *Tai Chi*. "Nature" as an expression appears first in *The Book of Change*, which maintains that *Tai Chi* generated Heaven and the earth, which, in turn, generated all things. In the Song Dynasty, it was developed by the then neo-Confucianists into the infinite Way of Heaven, thus becoming the perfect order that all things follow.
>
> Cosmologically, Nature is the hotbed for Heaven and the earth and all things. From the perspective of value theory, Nature is the process of all creations and evolutions and it generates different hierarchies of value such as forms of beauty and virtues and perfection achievable with the guidance of truth.
>
> The Chinese prefer *ziran* (Nature) to *yuzhou* (宇宙, or universe or cosmos) whatever the difficulty. The reasons are as follows:
>
> The first reason is that "The nature (of man) having been completed, and being continually preserved, it is the gate of all good courses and righteousness." (*The Book of Change. Xi Ci*) In Chinese philosophy, Nature in the sense of the natural world and human nature are identical, so this expression mirrors the oneness of Nature and Man.
>
> The second reason is that poetically inclined, the Chinese tend to "humanize" nature. A quote from Lao Tzu is a good case in point. Lao Tzu says, "That which was the beginning of all things under Heaven we may speak of as the 'mother' of all things. He who apprehends the mother thereby knows the sons. And he who has known the sons will hold all the tighter to the mother, and to the end of his days suffer no harm." Nature is to man what the mother is to the son. The Nature-man relationship never disappears simply because man may be detached from Nature as the son may be detached from his mother.
>
> The third reason is that in terms of hierarchy, Nature unifies Heaven and the earth and man, and combines all existences into a harmonious movement as a tribute to the wonderful universe. (Fang 128–129)

Fang Dongmei highly praises the classical Chinese conception of Nature as a generative, vital, free, harmonious, perfect whole unifying Heaven and man to make it an antithesis to the material, mechanical, and simplistic conception of nature as well as to the political, economic, and personality structures on the basis of such a conception. Unlike Qian Mu's rationalism, Fang keeps "reason" at arm's length and even blames the neo-Confucianists of the Song Dynasty for their "stubborn clinging to reason," and "violation against Nature both in the sense of humanity and the natural world." He points out on many occasions that "The Song neo-Confucianists cling to ill-grounded theories and annihilated virtuous human desires, tastes, feelings and emotions. Their philosophy cannot be juxtaposed with literature and arts and open cultural spirit; otherwise, a distorted philosophy system may take roots" (Fang 507–508). In order to correct the one-sidedness and absoluteness of "reason," which impairs human nature, Fang argues, "People should draw on the spirit of Daoism," since "the spirit of Daoism can be unified with Confucianism as they once were in the pre-Qin times" (508). Fang finds that the Confucianism coincides with Daoism on the idea that Heaven and the earth and man were all generated, and the ten thousand things (all creations) and the self are one. Gu Xiang's substitution of "the law of things" for "the Way of Heaven" is considered by Qiang Mu as "a major gift to Chinese thought." On the contrary, it is viewed by Fang as "the gravest of mistakes," because Qian is committed to

China's modernization, but Fang is committed to offering oriental resources for more profound reflections on modernity.

As one of the spearheads of modern neo-Confucianism, Fang believes that the valuation of "non-being" in Daoism is the path whereby the mainstream Western philosophy may be surpassed. He argues:

> The Daoist approach to philosophy is to proceed from non-being rather than being; in other words, it starts on a meontological level instead of a merely ontological level. The superficial being should be traced all the way to its essential non-being. In Greek ontology, the absolute being is to be understood via the relative being, and this, in the eyes of Daoists, still a clinging to being. However, the ultimate truth can only be revealed through negation. (257)

Those who favor "being" tend to emphasize utilitarianism and the endeavor to empower the nation; those we value "non-being," emptiness and transcendence, may better safeguard their personal purity, freedom, and dignity. In essence, being and non-being, as opposites of the unity, capture the two mutually complementary aspects of social activities and personal life. Eventually, the pro-being ones believe that man can conquer Nature, which is an object for them; the pro-non-being ones aspire to the oneness of Nature and man and to the salvation of personal existence. The former views earthly possessions as the greatest of blessings; the latter makes poetic dwelling their worthiest of pursuits. In fact, Qian Mu does not deny this. In his criticism of Daoism as "impractical," he points out that "When applied to arts, Daoist conceptions are truly transcendental." It is safe to say that Qian Mu, a critic of Daoism, still values Daoism, though he is committed to creating a better life instead of higher arts through Daoism.

The connection between art and life is interwoven with the perception of Nature. Fang Dongmei, again, among other modern Chinese scholars, expounds the spirit of Chinese philosophy and art in a thought-provoking way. He points out that classical Chinese philosophy and cosmology contain universal value of arts; for instance, Zhuangzi's idea that "being a saint means trying to understand the law of things through the beauty of Heaven and the earth" embodies the Chinese spiritualism; the Chinese must become artists before they become thinkers, because their perception of the nature of beauty usually precedes that of other things. For this reason, the Chinese are better at artistic creation, especially poetry. Even Chinese philosophical wisdom develops through an immersion in arts. Chinese philosophy and poetry are of the same roots historically, and both of them point to the unity of Nature and life. Fang says, "In Chinese arts in general, there is vitality in the form of freedom and eternal aura. It symbolizes a eulogy for and appreciation of the eternal magic touch of the universe which generates and nurtures all things" (373). However, Fang also states that "Without a decent understanding of Daoism, one would have difficulty appreciating the significance and subtlety of Chinese arts such as poetry and paintings" (368).

More often than not, Fang, rooted in Confucianism though, tries to trace the origin of Chinese philosophy to *The Book of Change*. He infuses the personality of Nature to Confucian "virtue," thus creating "the virtue of Nature's soul" (413). In such connectedness of Nature and man, Chinese poets are always able to "feel and contemplate Nature's transformation and their own oneness with Nature" (381), thus making possible poetic dwelling. Fang's statement offers some trustworthy insight into Tao's poetic philosophizing and life.

It seems that in the modern evolutions of Confucianism, Daoist spirit has penetrated the whole of Confucianism and reformed the latter's teachings and even stance though Confucianism is still predominant in studies of traditional Chinese thought. In terms of attitudes to Nature, if Qian Mu and Fung Yu-lan perhaps try to save Confucianism by virtue of Western Rationalism and pragmatism after the Enlightenment Movement, and therefore are cautious about Daoism, Fang Dongmei, informed by Bergson's life philosophy, Freud's psychoanalysis, and Whitehead's process philosophy, is open to Daoism, especially the Daoist idea that "Dao follows the so-of-itself," and to critical of instrumental rationality, thus becoming part of the Chinese reflections on modernity.

When Tu Weiming (1940–) emerged as "the third generation heir to modern neo-Confucianism," Confucianism faced another major reform in a context of world-sweeping ecological crisis that jeopardizes each and every nation's and individual's subsistence. Unlike Qian Mu, known for his rationality, and Fang Dongmei, known for his life philosophy, Tu has adopted an eco-philosophical approach to the reform of neo-Confucianism. Naturally, Daoist natural philosophy, which contains ecological and even spiritual-ecological significance, has risen to become a drive for Tu's endeavor to reform Confucianism and to make Confucianism part of the resources for reshaping the post-modern society. Therefore, Tu's ecologically inclined neo-Confucianism has a post-modernist touch.

Above all, Tu emphasizes that social development must be "reoriented" to better handle the global ecological crisis, saying, "There has to be a fundamental and urgent change to the relationship between Nature and man both in theory and in practice for the sake of human subsistence"(Tu "DI" 182). As is known, "ecology," which emerged originally as "natural science," began its "humanist turn" partly because of the voice of Rachel Carson (1907–1964) in the 1950s and now it has found its way into many fields of humanities and social sciences such as sociology, politics, economics, ethical philosophy, the science of law, philosophy, aesthetics, and literary criticism, where the relationship between Nature and man is inevitably re-examined. In such a context, Tu proposes "the ecological turn of Confucianism" in explicit terms, arguing that this is a brand-new world outlook featuring "the organic and interactive unity of cosmos and human society," and that "in terms of Confucianism re-evaluation, such a world view marks the cultural turn of Confucianism in that it emphasizes the interaction between the earth and man" ("DI" 183). He argues that "Confucianism must go beyond anthropocentrism" and the "human-centeredness" ("DI" 2), a basic teaching of Confucianism, must be changed. To enhance his argument, Tu cites the observations of Xiong Shili (1885–1968) and Liang Shuming (1893–1988) on the Nature-man relationship, especially the former's "Nature vitality theory" based on *The Book of Change,* which preceded the ecological turn of neo-Confucianism.

Drawing on the experience of modern neo-Confucianism over the past century or so, Tu's series of lectures in the early twentieth century are a constant reminder of the necessity of self-reflections. Tu found that since the May Fourth Movement, Confucianism, downsized by pro-Western scholars, has begun to incline toward Western thought. He claims that ridiculously, the evaluation of Confucianism should have become whether Confucianism conforms to modernization in the Western sense of the definition, and this is considered by Tu as "the modernist turn of Confucianism," which, consequently, has rendered Confucianism scientificized, instrumental, secularized, and simplistic because "the original Confucian language has been reconstructed fundamentally, and therefore, it is no longer a language about faith, but a language about instrumental rationalism, economic benefits, political expediency and social governance" (217). Such Confucianism is adverse to ecology instead of getting any closer to it. At present, Tu is committed to leading modernist Confucianism back to the naturalistic and ecological track and to re-energizing the conception of the oneness of Heaven and man. This, for him, will mark a rude awakening of neo-Confucianism from the trance of modernism, which will contribute to the reconstruction of a brand-new outlook of the world and life as well. For Tu, the reconstruction of a post-modern ecological world and life view necessitates the absorption of vitality from traditional Chinese thought, especially the conception of the earth as a "vital organic community of life." Such a conception can be Daoist or Confucian or a mixture of both and others. Tu's vital organic community of life as such is akin to Nature's transformation in classical Chinese philosophical term, whose essence is "*qi,*" the vital life-force pervading all, both as substance and energy, life and vitality, being and non-being, and both physical and spiritual. It generated all things as per classical Chinese philosophy, and the generation of Nature is Nature per se. Tu points out that human beings themselves are part of the Heavenly Way just as rivers and mountains are the legitimate beings of Nature's transformation or the ultimate results of the flow of *Qi* as such (Tu "CB" 88–93).

One of Tu's major contributions to Confucianism lies in his reinterpretation of classical Chinese natural philosophy from a modern ecological point of view, and, more importantly, in his deconstruction of the dualistic modern thought pattern such as matter vs spirit, subject vs object, and in his breaking down of the barrier between Confucian being and Daoist Non-being, thus presenting a Chinese version of cosmology, which will eventually benefit the reconciliation or harmony between Nature and man in a post-modern era.

2.4 "Knowing the Bright but Cleaving to the Dark" and "Tao's Names"

"Knowing the bright but cleaving to the dark" appears in *Lao Tzu* (Chapter 28). Arthur Waley's translation of the chapter is as follows:

> He who knows the males, yet cleaves to what is female
> Because like a ravine, receiving all things under heaven,
> And being such a ravine
> He knows all the time a power that he never calls upon in vain.

This is returning to the state of infancy.
He who knows the white, (yet cleaves to the black
Becomes the standard by which all things are tested;
And being such a standard
He has all the time a power that never errs,
He returns to the Limitless.
He who knows glory,) yet cleaves to ignominy
Become like a valley that receives into it all things under heaven,
And being such a valley
He has all the time a power that suffices;
He returns to the state of the Uncarved Block.
Now when a block is sawed up it is made into implements;
But when the Sage uses it, it becomes Chief of all Ministers.
Truly, "The greatest carver does the least cutting."

Gao Ming (1926–) states that this chapter boils down to three layers of meanings, namely "knowing the male but cleaving to what is female," "knowing glory but cleaving to ignominy," and "knowing the bright but cleaving to the dark," which correspond to three dualities, respectively, which, in turn, are the strong and the weak, the noble and the humble, and the visible and the invisible. The real nexus of the statement herein is "knowing the bright but cleaving to the dark," which, ultimately, is a symbol of the limitless or *Dao*. Informed by Daoist spirit, Gao compares the infant's innocence to the original state of human beings, the uncared block to the original state of trees, and the limitless to the creation, or cosmology in the ontological sense. Zhang Zhiyang (1940–) believes that Chapter 28 is the cornerstone for the whole of *Lao Tzu*, and that "knowing the bright but cleaving to the dark" is the culmination of the great Daoist Way as well as an epitome of Daoist spirit in general. By the way, the famous *Tai Chi* diagram is represented as a mutually becoming and mutually chasing white (the bright) and black (the dark).

It is no surprise that Niels Bohr (1885–1962), the Nobel-winning Copenhagen quantum physicist had the diagram inscribed on his medal. In fact, the Daoist mutually complementing and mutually becoming dualities such as white and black, the bright and the dark, the visible and the invisible, and the strong and the weak, the hard and the tender, and the heavy and the light have long been an attraction for scholars. A new theory of cosmology believes that the cosmos after the big bang was just like a butterfly having two wings which spread in opposite directions, and what we know as the universe today is merely one of the two wings because the other wing is still hidden in the dark. In one of his lectures in Shanghai, Tsung-Dao Lee (1926–), a Nobel laureate for physics, says, "Surprisingly, the mass of the matter that is known to us only accounts for 5 % of the total mass of the universe, and the dark matter which is 'invisible' now accounts for 95 % of the universe's mass. In the depth of the universe, there may exist heavy particles as well as light ones. The 'dark energy' of the universe may be 14 times more than the energy that is known."[3]

[3]For more details, see Wenhui Bao, 2009–06–16.

As regards the domain of spirituality or in the individual's inner cosmos, "knowing the bright but cleaving to the dark" reminds people of the iceberg theory of Freudian psychoanalysis: sub-consciousness, like the most of the iceberg, remains under the water; and collective unconsciousness is even deeper under the water. Jung tries to explain the significance of "knowing the bright but cleaving to the dark" to the Western world in his own terms, saying that what is really significant in psychology is usually hidden in the dark and that in the spiritual life of a nation, the dark can evoke the bright for people. This may help to explain the increasing popularity of Jung in Western scholarship and among the general public.

"Knowing the bright but cleaving to the dark" and "knowing the visible but cleaving to the hidden," therefore, seem to be a path leading to the ultimate significance of Nature as well as to the innermost depth of human soul. Unsurprisingly, a study of Tao Yuanming's name would reveal the significance of "knowing the bright but cleaving to the dark" that it contains.

Admittedly, Tao's name has remained a riddle. According to Zhu Ziqing (1898–1948), there are up to ten allegations about Tao's given name and style name. Zhu himself believes that Yuanming (渊明) is Tao's given name, and "yuanliang" (元亮) his style name, which Tao himself changed into "qian" (潜) when the Jin Dynasty was replaced by the Song because of his "strong Daoist inclinations" (Zhu 458).

Like Zhu Ziqing, Yuan Xingpei (1936-) intends to decipher the significance of Tao's name from the dimensions of classical Chinese philosophy by citing *The Book of Change*:

> As regards the connectedness between "*Yuanming*" (meaning literally "abyss bright"), "*Yuanliang*" (meaning "bright") and "Qian" (meaning "hidden" or "seclusion"), *The Book of Change* states that "As the dragon is hidden in deep water, it is not the time for action. The dragon may also be in the abyss." According to *Guangya*, *qian* means 'hidden', and according to *The Origin of Chinese Characters*, 'hidden' means 'cover'. Both 'hidden' and 'cover' point to 'darkness' as opposed to brightness. According to *The Origin of Chinese Characters* again, *liang* as in *Yuanliang* means brightness, an antithesis of *qian* ('hidden'), and, at the same time, a synonym for *ming* (brightness). (Zhu ZQ 237)

Semantically and epistemologically, Yuan's interpretation conforms to the most common structure of the names of literati of antiquity. It may be an accurate deciphering of the significance of Tao's names. Gao Heng (1900–1986) explains that "The dragon is hidden in deep water, which is the right place for it. When a man finds his comfort zone, he has little to complain about" (161). This explanation completely agrees with Tao's "refusal to change for the sake of earthly achievements and fame," his "return to the rustic life," and his "going back to Nature"; therefore, it should be valid.

I want to add that if Tao Yuanming changed his name at all, it must have happened after he decided to resign from his official post because he viewed the political career as a "cage" or a "net." "Yuan" as in "yuanming" may be derived from *The Book of Poetry*, which contains such a line: "Fish may hide in the dark abyss or bright shallow waters near islets." In fact, "abyss" appears many times in Lao Tzu, and the Way is often compared to the abyss. Without doubt of any, the dark abyss, a haven for fish, is much safer than the bright shoals. Such an explanation may help better understand the connection between Tao's names and his life philosophy. Ultimately, Tao's names, though multiple, boil down to two antithetical aspects, namely brightness and darkness. What Tao Qian or Tao Yuanming means "a ray of brightness hiddens in the depth of the dark," which is reminiscent of the Daoist teaching of "knowing the bright but cleaving to the dark."

Though the exact expression "knowing the bright but cleaving to the dark" is not found in Tao's poetry and prose, yet this, for Tao, is not only a generative pattern of things, but also a way of beautiful existence. Tao Yuanming's observation of this way is expressed in his self-naming, and, more importantly, in the easy graceful balance he keeps between glory and humbleness, wealth and poverty, drunkenness and sobriety, the past and the present, and words and significance. This is only possible because he lives by "knowing the bright but cleaving to the dark."

2.5 "Knowing the Bright but Cleaving to the Dark" and Tao's Wisdom of Life

A glimpse of the world history would reveal that, almost unexceptionally, great writers and poets tend to identify themselves with the core cultural and spiritual legacy of the nations they belong to, thus epitomizing their nations' wisdom of life and becoming poetic avatars of the nations' souls. Tao Yuanming is no exception. He keeps to the principle of "knowing the bright but cleaving to the dark" in life and crystalizes it in his writings, the evidence of which abounds.

(1) Returning: Official Career and Hermitage

Returning to the country life was an important event, if not the most important one, in Tao's life. Tao confesses, "A man who stays away from worldly quest, I never seek for office east and west" (187).

From the age of 30–40 years old, Tao came into a political career several times, first as Consultant for a general and then as Head of Pengze Prefecture, without achieving any political feat. It is said that Tao, at the age of forty, unwilling to ingratiate himself with the superior inspectors, abandoned his post and went back to his ancestral home and lived a *tianyuan* life ever after.

Social factors must have contributed to Tao's tenacious determination to retreat. For him, in a context where the political career was a cage, the networking, and treachery of the upper class was rampant, political corruption was depressing, and even life itself was threatened, abandoning the official post meant "returning" before going astray, or "becoming fish hidden in the depth of the abyss."

However, ultimately, Tao retreated because he followed his inner voice, a voice growing louder as each day passed by, as he confesses in his "Homeward ho!": A few days into my official post, I began to think about leaving, because I am Nature-bound instead of being affectionate. He describes his *tianyuan* life this way:

It is a pleasure for me to pace the yard,
With the state closely shut all day long.
With a staff in hand, I walk to and fro,
Rasing my head to look afar off and on.
The careless clouds float from behind the hills;
The weary birds know that they should return.
When the dim sun is about to set in the west,
I fondle a solitary pine and linger around it. …
I may ascend the eastern heights to sing a song,
Or sit by a clear stream to write a poem.
I shall follow the natural cause and end my life in time;
With Heaven's decree in mind, what else am I to doubt? (Tao 245–247)

In Tao's writings, returning to the *tianyuan* means returning to Nature, to the origin of life, to the soul's dwelling, and to the poetic. Glen Love (1932–) once elaborates upon the significance of the *tianyuan* to literature and humanity from an eco-critical point of view, saying, "The lasting appeal of pastoral is a testament to our instinctive or mythic sense of ourselves as creatures of natural origins, those who must return periodically to the earth for the rootholds of sanity somehow denied us by civilization" (225). Unfortunately, the root reason for Tao's retreat, which is his Nature-rootedness, is often ignored.

For Tao, to shun the hustle and bustle of earthly affairs is to "know the bright," and to return to the *tianyuan* and hermitage is to "cleave to the dark" or to safeguard the purity of the soul. Lao Tzu says, "(And being such a ravine) He knows all the time a power that he never calls upon in vain. This is returning to the state of infancy" (Chapter 28). On another occasion, Lao Tzu says, "He who knows glory yet cleaves to ignominy becomes like a valley that receives into it all things under heaven; and being such a valley, he has all the time a power that suffices; He returns to the state of the Uncarved Block" (Chapter 28). Here in this case, the uncarved block means the simple, original, unwrought state of being, which is the highest achievement of the enlightened Daoists. Tao's returning to the *tianyuan* is a lived experience of "returning to the state of the uncarved block."

In the same light, Heidegger highly values Hölderlin's *Homecoming*. For Heidegger, the homeland is the origin and the root of the soul, which, in turn, has to dwell in a haven as the tree must take roots in the earth. Even the tree follows the principle of "knowing the bright but cleaving to the dark": Its twigs and foliage are exposed to light, but its roots remain in the dark earth. Lao Tzu says, "See, all things howsoever they flourish return to the root from which they grew. This return to the root is called Quietness; quietness is called submission to Fate; what has submitted to Fate has become part of the always so" (Chapter 16). It is in this sense that Tao's returning to the *tianyuan* is "returning to the root," almost identical with Heidegger's aspiration to returning to the origin, where the true poetic dwelling is possible. In fact, the ultimate significance of Tao Yuaming's returning to the *tianyuan* lies herein.

(2) The Peach-blossom Springs: Reality and Vision

Tao's *The Peach-blossom Springs*, one of the best of Chinese writings, ever since its appearance, has remained a resource of inspiration, and an Arcadia, or a Shangri-La, or a land of promise for later men of letters and general readers alike in China.

Unlike the real society, Tao's Peach-blossom Springs is a tranquil magic other-worldly world which he himself terms "a world of immortals," which, in essence, is an idealized world of freedom and simplicity. "The Peach-blossom Springs" has become a synonym for romanticist spirit and Utopian imagination. Liang Qichao points out that Tao, in his writings, envisions an ideal social organization, and therefore names it "the oriental Utopia" (25). Zhu Guangqian has similar remarks, saying that at a time of social chaos, and witnessing people's sufferings caused by what were called institutional rules and regulations, "Tao realized that the nation's lifeline was ultimately farming and only farming-related engagements and entertainments can provide true pleasure of life, so he rested his longings on a simple Utopia like the Peach-blossom Springs he envisaged" (Zhu GQ 217–218). It is worth noting that the fact both Liang and Zhu use descriptors such as "oriental" and "simple" imply that they do not think the orthodox "Western Utopia," designed by Sir Thomas More (1478–1535), is as simple as it is because it has to be based on reason and knowledge, governed by rules, supported by affluence, and realized by the development and transformation of nature. A comparison of Tao's Peach-blossom Springs and the Western Utopia would indicate that they have little in common except for the rough likenesses shared by all literary imaginations. Unlike the bright, progressive, and promising Western Utopia, Tao's Peach-blossom Springs is of a retreat and hermitage type.

Only after the many mirages and visions such as Plato's "Republic," Bacon's "New Atlantis," More's "Utopia," Tommaso Campanella's "City of the Sun," James Harrington's "Oceana," Robert Owen's "New Harmony," and Etienne Cabet's "Ecarian Community" have been "realized" in the real world as steam engines, inner combustion engines, auto assembly line, nuclear submarines, space shuttles, robots, cloned dogs, genetically modified food, skyscrapers, express ways, supermarkets, transnational banks, and various kinds of charters, forums, committees, and boards of trustees, did people realize that happiness they have been trying to seek has never come as promised, and that, even worse, the realized "Utopia" is just like a shining edifice which in essence is a cage. As a result, new thinkers begin to emerge on the land of so-called Utopia and throw their spears at the orthodox "Utopia." As anti-Utopianists, they satirize and criticize "Utopia" of the Enlightenment style and express their deep concerns about the deteriorating ecology, society, and even civilization, consciously or otherwise, thus ushering the world's Utopian movement into a post-modernist era. In this context, Tao's Peach-blossom Springs, in retrospect, now is shimmering as a ray of light of salvation. Tao's simple oriental "Utopia" coincides with the "anti-Utopia" envisaged by Yevgeny Zamyatin (1884–1937), Aldous Huxley (1894–1963), and George Orwell (1903–1950) in that what they have presented is exactly the antithesis of the orthodox "Utopia."

(3) Drinking: Sobriety and drunkenness

Half of Tao's writings relate to drinking. Considering his confession in his lines like "Keeping my shadow company, I drink every evening. I know not that I am already drunk" and "After hundreds of years, who would know about my glory or ignominy. The only regret would be that I have not drunk much in this life," readers are under the impression that Tao does not spend a single day without drinking until the last grain of sand of his life runs out. In public opinion, drinking, especially binge drinking, is not infamous, but when it is associated with men of letters and poets in particular, drinking can mean a very different thing. Tao may be the best epitome of the poets of whose life drinking is an inseparable part.

Regarding why poets tend to be positively indulgent in drinking, opinions are divided, especially on Tao's case.

In the corrupt age of the Wei and Jin, scholars had a difficult time, and therefore, they tended to drown their sorrows in wine, and in so doing, they would forget the earthly world, their sentimentality, their sufferings, and even themselves, and thus became united with Nature. Hu Bugui (1906–1957) remarks that "Drinking is the gateway for Tao to return to the fundamental and the true" (127). This is a valid observation. In other words, drinking is the means by which Tao enters into non-being from being, into the dark from the bright, and into otherworldliness from this worldliness. In fact, Tao vividly expresses a similar opinion in his *Drinking XIII* through a contrast of the drunken man and the sober man. Tao writes:

To stay together as one's closest guest,
With counter aims the other shows his zest.
The one is oft in a drunken state
While the other is sober and awake.
The two of them would laugh with cheers,
But never give each other listening ears.
The sober man is foolish in disguise;
The drunken man is proud but much more wise.
Please bring my message to the drinking man:
By candlelight keep drinking while you can. (119)

In his poem, the intoxicated man has much more wisdom than the sober one, and this, according to Ye Jiaying (1924–), what underpins the poem is Daoist thought. In Daoism, there are teachings on "depend-on" (external underpinning for gratification) and "non-dependent-on" (internal or self-underpinning for self-nourishment or gratification). The sober man is calculative, indecisive, and penny-wise, but the drunken man is pound-wise by listening to his inner voice and living a seemingly vegetative life. Such poetic philosophy echoes in spirit with Lao Tzu's "knowing the bright but cleaving to the dark" and his statement that "The way out into the light often looks dark; the way that goes ahead often looks as if it went back" (Chapter 41). Obviously, the valuation of drunkenness over sobriety and that of non-being over being is another form of "knowing the bright but cleaving to the dark" on the spiritual level.

Heidegger points out in his elucidations on Hölderlin's "Remembrance" that poetic intoxication (*Trunkenheit*) is different from ordinary drunkenness: In the former case, poets, in their intoxication, go beyond the everyday logical cognition and elevate their souls closer to the origin of being, so for them, intoxication is a true state of being as poetry is the best means of contemplation. Unlike anesthesia, intoxication, for Heidegger, is a sublime form of attunement, and it brings poets to lucidity. He says, "… wherein the depths of concealment are opened up and darkness appears as the sister of clarity …" (143). This may be what Tao means by "the significance in drinking." It is safe to say that poetic intoxication in Heidegger's term is exactly what is known as "Tao's intoxication."

(4) Elegy: Life and Death

Regarding life and death, Tao does not believe in Daoist immortalization or Buddhist Samsara (cycle of life). His philosophicality about death can be attributed to his idea that life is but a lodge for the journey, saying that life is about transformation; it eventually vanishes into non-being. For Tao, death means the final oneness with the earth and Nature's transformation, and with the Way or non-being. Life and death for Tao also fall under the categories of "the bright" and "the dark," respectively: Life means "the bright," which is finite and ephemeral; death means "the dark," which is infinite and eternal. Tao's view of death as "returning" is based on his practice of "knowing the bright but cleaving to the dark," thus rising from transiency to eternity.

The Daoist conception of life and death is based on natural philosophy. For Zhuangzi, life and death are just like the shift of day and night or the rotation of seasons, whose essence is the amassing and dispersal of *qi*, or life-force, which, in turn, is part of Nature's transformation. For Zhuangzi, life is followed by death, and death is the beginning of life, who knows the law; life is generated by the amassing of *qi*, as death caused by the dispersal of it. Tao also has this idea: Since life comes from Nature and to Nature it returns, which is a law independent of man's will, why should one trouble oneself with the way to go against the Way? He says, "The human life is like a magic show; to nothingness it will eventually go" (57). In fact, Tao's being comfortable with and philosophical about life and death infused into his life, extra pleasure and ease, which he would otherwise have missed.

Prior to his long departure, Tao composed himself an elegy, saying that death is no big deal; it just means the oneness of the body with the mountain. In his "A Funeral Oration for myself," he writes:

> I am about to stop my earthly sojourn and return to my eternal underground residence… The expansive earth and the boundless sky have given birth to everything, including me as a human being. … As I never overworked myself, I was filled with ease and comfort. To obey the laws of the heaven and follow the natural course of events – that is the way I spent my life. … Now I am to leave this world without any regret, for I have attained my ideal of living in the countryside. Now that I am to die a natural death at my old age, what is there for me to linger on? (275–277)

Untroubled by life and death and comfortable with Nature's transformation, Tao moves toward spiritual freedom and thus has discovered a rare gateway to the highest state of poetic dwelling.

2.6 Tao's "Knowing the Bright but Cleaving to the Dark" and Heidegger's Philosophy

Coincidentally, among the Chinese philosophical works that Heidegger mentions is Chapter 28 of *Lao Tzu*, which, as one can judge, has influenced Heidegger's philosophy profoundly.

During World War II, Heidegger translated eight chapters of *Lao Tzu* in collaboration with Xiao Shiyi (1911–1986). They translated what is known as "knowing the white but cleaving to the dark" into "He who knows his brightness veils himself in his darkness" (Poggeler 63). Heidegger believes that "Mortal thinking must let itself down into the dark depths of the well to see the star by day" (Poggeler 62). Poggeler says that Heidegger "did not mention the 'star' on a momentary impulse: while since Plato the sun has been an image for divine reason, which bathes everything in its light without darkness. The star rises alone for us out of darkness and its mysterious depths. But Heidegger seeks in Lao Tzu the trace of the most ancient thoughts only in the service of leading that which determines our history back to its hidden origin" (Poggeler 62). "Knowing the bright but cleaving to the dark" must have been inscribed in the mind of Heidegger of old age because according to his will, inscribed on his headstone was not a typical crucifix, but rather a shining star which implies that "the star rises alone for us out of the darkness and its mysterious depths."

Interestingly, Heidegger, as a lone star immersed in the dark after he ceased to be, seems to be a footnote to Tao Yuanming's name. He and Tao are somehow connected by a spiritual bond, which is "knowing the bright but cleaving to the dark," as Zhang Zhiyang (1940–) argues, "As a matter of fact, Heidegger's thought can be described as knowing '*sein*' but cleaving to '*Dasein*' and 'existence',[4] modeling on Lao Tzu's paradigm of 'knowing the bright but cleaving to the dark'. We should have realized earlier that Heidegger, on its way back to the Pre-Socrates philosophy of Greece, echoed Tao Yuanming across time and space" (Zhang ZY 7–14).

Poggeler describes the connection between Heidegger's thought in his later years and Daoist philosophy this way:

> When Lao-Tzu saw that his country was declining irrevocably, he left the archive at the court of Chou in order to wander across the border. As Heidegger engaged the Lao-tzu, he on his side left the archive in which the Western tradition was articulated. The return to the ground of metaphysics was also a way to one's own origin, and yet this origin, which was supposed to have left substance-thinking behind, was, in turn, understood again as a pre-given, quasi-substantial beginning. (66)

Poggeler believes that Heidegger's philosophical return is to liberate modern people from the dominant power of Plato over the past two millennia. Ultimately, Heidegger's path of "return," like that of Tao, also leads to Nature and freedom.

[4]The German word Dasein is sometimes translated as "being-there" or "being-here"; the nature of Dasein's being is described by Heidegger as "existence." Tr.

Heidegger meticulously elucidates Hölderlin's description of the "dark light" in "Remembrance." According to Poggeler, this is another example of Heidegger's drawing on "knowing the bright but cleaving to the dark," saying, "Heidegger supplements this phrase of Hölderlin's with Lao Tzu's line about the wise: he who knows his brightness veils himself in his darkness" (66).

To put the reader better in the context, Hölderlin's stanza in question is cited herein:

> But someone hand me
> the fragrant cup,
> full of dark light,
> that I may rest.
> It would be sweet
> to slumber in the shadows.
> It isn't good
> to stay mindless
> with human thoughts.
> On the other hand, conversation
> is also good: to speak
> the thoughts of the heart,
> and to hear much of days of love,
> and of deeds that occur.

Centered around "dark light," Heidegger's elucidation is as follows:

> The wine is named the dark light. Thus at the same time the poet asks for the light and for the brightness which contributes to clarity. But the dark light, in turn, cancels out the clarity, for the light and the dark are in conflict. Or so it seems to be for that kind of thinking which is exhausted in calculating with objects. The poet, of course, sees an illumination which comes to appearance through its darkness. The dark light does not deny clarity; rather, it is the excess of brightness which, the greater it is, denies sight all the more decisively. The all-too-flaming fire does not just blind the eyes; rather, its excessive brightness also engulfs everything that shows itself and is darker than darkness itself. (141)

To some degree, Heidegger is more like presenting his own poetics, especially the relationship between brightness and darkness than elucidating Hölderlin's poetry. Lao Tzu's "knowing the bright but cleaving to the dark" and his statement that "the way out into the light often looks dark," and that "great white often looks black" can be considered as a footnote to "dark light" as in Hölderlin's "Remembrance," where darkness and brightness are united harmoniously. If "brightness" represents "being," or things out there, then "darkness" corresponds to "*Dasein,*" or the root of all things. Only in poetry can the origin of all things be revealed via "dark light." What Hölderlin drank was wine, and Hölderlin might have never had a sip of Chinese liquor, in which water and fire are perfectly blended as "brightness" and "darkness" are. Like the mutually generative water and fire or brightness and darkness, the future and the past, sobriety and drunkenness are also integrated as one in Hölderlin's poetry, as can be seen from Heidegger statement as follows:

> The darkness preserves in the light the fullness of what it has to bestow in its shining appearance. The dark light of the wine does not take away awareness; rather, it lets one's meditation pass beyond that mere illusion of clarity which is possessed by everything calculable and shallow, climbing higher and higher toward the loftiness and nearness of the highest one. (142)

In his later years, Heidegger often compares human beings to "plants." He quotes Johann Peter Hebel in his "Memorial Address to Konrad Keeutzner," saying, "We are like plants, which-whether we like to admit it or not—must rise out of the earth in order to bloom in the ether and bear fruit" (Sun 1241). Taking roots in the earth and reaching out into the sky—this is an example of Heideggerian "knowing the bright but cleaving to the dark."

Informed by Oriental philosophy, Heidegger bases his contemplation of "being" on the distinction between "being" and "Dasein." For Heidegger, modern philosophy can only deal with "being," whereas it should address "non-being" as complete transcendence over "being."

As to what "non-being" is actually is, Heidegger's answer is as follows:

> The no-thing is not an object or anything that is. The no-thing does not show up either for itself or alongside things as if it were an add-on. Rather, the no-thing makes possible the appearance of meaningful things, as such, for human being. The no-thing is not just the opposite of things; it is essential to their very emergence. The repelling action of the no-thing takes place in the very meaningfulness of things. (Sun 146)

The above-cited elaboration, vague and esoteric, is similar to Lao Tzu's discourse. Like Lao Tzu's "non-being," Heidegger's Nichts (no-thing, non-being) transcends the being of all things. What sets Nichts apart from Lao Tzu's "non-being" is that in Heidegger, "Nichts" is the background and destination of all things, and therefore "being" is built into "no-thing"; for Lao Tzu, "being" is generated out of "non-being," which, in turn, is the origin of all things. Ultimately, Nichts and Lao Tzu's "non-being" serve the same purpose: to transcend over reality and return to or guard the pure unwrought state of "being." In this sense, it is safe to say that Heidegger's distinction between "being" and "no-thing" is a Western version of Lao Tzu's "knowing the bright but cleaving to the dark," the bright representing being or the visible, motion, progress, unveiled reality, and things out there, and the dark signifying no-thing, or the invisible, motionlessness, regression, hidden reality, and the origin of things. In addition, what binds Heidegger and Lao Tzu across time and space is that for them, their "Nichts" and "non-being" are like unfathomable abysses and that poets' and philosophers' contemplation is a ray of hope in the depth of the abysses.

How can one feel the "being" of "non-being" or "Nichts" and somehow enter into the state of the latter since "non-being," as the original unwrought state of "being" which beggars everyday expression, is not a subject of scientific cognition or logical analysis?

Originally, Heidegger discovers this path: "angst" (fear) informs "Nichts," and "death" makes people face the disappearance of all things out there; therefore, "Nichts" is revealed. Angst is unspeakable overwhelming "fear" or terror under which all cleavings and underpinnings in the real world give way, and, therefore, one enters into a vague aloofness and permeating vastness. It is now that one can feel the impending "Nichts" and therefore feel self-existence more keenly, as Heidegger says, "Without the original revelation of the no-thing, there is no selfhood and no freedom" (Sun 146). Zhang Shiying offers an interpretation to Heidegger's statement, saying, "Man, in everyday life, is led by the nose by institutions and norms. However, in the face of the impending death, all earthly concerns vanish since death points to Nichts. As a result, the original true state of man appears. Only at this point in time can one feel and discover one's 'ontological self', thus reaching a state of absolute freedom. So, Heidegger's philosophy about Dasein and Nichts is essentially about the return to 'origin' and freedom" (Zhang SY 373–374).

However, Heidegger's path of achieving Nichts through angst aroused by impending death seems to be too "logical" in his argumentation and therefore hardly applicable.

In contrast, the Daoist return to the "origin" and "non-being" is not to be achieved through the "from-death-to-life" path taken by Heidegger. Instead, it is realized through the purgation of all concerns, the purification of the heart, the utter oblivion of the self and things out there, stillness, the forsaking of words and wisdom, the resting of one's mind in tranquility and solitude, the transcendence over time and space, and the breaking away from the yoke of "things." In doing so, transcendence or absolute freedom ensues.

Zhuangzi describes "heavenly music," the highest form of music, as "permeating and encompassing though soundless and formless," whereby one reaches "non-being." It is said that Tao Yuanming had a Chinese zither or harp without strings, implying the Daoist idea of "the great sound being soundless." Tao must have understood that music dwells in the heart rather than in things.

Heidegger believes that those who can realize "Nichts" through angst are always the great and the courageous. In contrast, those Chinese who can reach "non-being" through stillness are usually those of the easy abandoned hermits, most of whom are poets. This is an example of the differences between the China and the West in terms of deep national traits. In ancient China, the absolute freedom that comes along with realized "non-being" is, at the same time, always a state of the arts since the achievers were predominantly poets.

Zhang Shiying also compares Heidegger's and Tao's poetic philosophizing centered about "death" and "non-being."

Heidegger believes that death makes possible man's realization of "Nichts"; Tao writes in his poem that "As part of the great transformation, Life will be reduced to non-being." Zhang says, "In Tao Yuanming, there is no lament over impending death, but only realized "non-being" as transcendence over wealth, fame, things out there, and life and death. Tao's philosophizing can be viewed as a chanting of Heidegger's transcendence philosophy" (Zhang SY 376). However, since "poetry can convey the true significance of 'being' better than philosophy does" (Zhang SY 376), Tao may have felt transcendence and "non-being" more keenly than Heidegger once did. As regards "non-being" and the way of realization, Zhang cites Tao's poetic line "a remote heart creates a remote dwelling," arguing that philosophically, poetic remoteness as such is "transcendence" in

Heidegger's term. Both "remoteness" and "transcendence" point to the detached attitude toward the world. Heidegger in his late years argues that poets should be evoked to realize philosophical transcendence. He might have no idea that one millennium ago, Tao Yuanming already demonstrated how it is possible. It seems that Heidegger has failed to connect into a whole his "contemplation on death" in his earlier life and later on his thought on poetry, but Tao's contemplation on death is governed by his poetic "remoteness of the heart." Tao writes:

> I pluck hedge-side chrysanthemums with pleasure
> And see the tranquil Southern Mount in leisure.
> The evening haze enshrouds it in fine weather
> While flocks of birds are flying home together.
> The view provides some veritable truth,
> But my defining words seem to me uncouth. (113)

"By 'truth', what is meant in Tao is the self-so of life, akin to 'true state of being' in Heidegger's term" (Zhang SY 376–367). This observation is accurate.

In his late years, Heidegger gave up on the "from angst to Nichts" path; instead, he rested his hope of returning to the true state of being on poetry and poets. In Heidegger, art in general and poetry in particular become the original way of the generation of "the truth of being." Thus, he gets closer to Taoist philosophy. Heidegger sates:

> Yet the poet, if he is a poet, does not describe the mere appearance of sky and earth. The poet calls, in the sights of the sky, that which in its very self-disclosure causes the appearance of that which conceals itself, and indeed as that which conceals itself. In the familiar appearances, the poet calls the alien as that to which the invisible imparts itself in order to remain what it is – unknown. The poet makes poetry only when he takes the measure, by saying the sights of heaven in such a way that he submits to its appearances as to the alien element to which the unknown god has 'yielded'. (Sun 476)

In this case, what is meant by Heidegger is that only the poet can be a liaison in between Heaven, the earth, God and man, a soul that goes into and comes out of being and non-being, and one that internalizes Nature and lets it radiate in the soul. This is reminiscent of a poetic line by Chang Jian (708–765) reading "The zither music brightens and deepens the river." Metaphorically, poets are the zither music that elevates and transforms things out there.

Lao Tzu says, "I alone seem to have lost everything. Mine is indeed the mind of a very idiot. So dull am I. The world is full of people that shine; I alone am dark" (Chapter 20). It means that the enlightened ones are those guarding the dark lonely. Similarly, Heidegger believes that the enlightened ones are those plunged into darkness by great light. They could be philosophers like Heidegger himself, or poets such as Hölderlin and Tao Yuanming.

Though Tao and Heidegger lived in different times and states and faced different issues, yet their philosophizing, especially poetic philosophizing, has much in common. That being said, what confronted Heidegger was the deep-rooted tradition of rationalism, the trauma inflicted by scientific development on ecology, and the aloofness or rootlessness of human spirit caused by the subversion of traditional values. In Tao's day, these problems had not surfaced. The issues that did confront Tao seem to be of a "lower" level, or, merely problems about how one deals with things and the world and how one can live with more freedom and originality and less confusion. Paradoxically, the seemingly "lower-level" issue has more universality and spontaneity, and therefore, it precedes other concerns.

Chapter 3
Tao Yuanming and Naturalistic Romanticism

Romanticism remains a complex issue, especially in China's literary criticism. It comes as no surprise that Isaiah Berlin (1909–1997) laments in his Roots of Romanticism that romanticism is a dangerous chaotic area just like a strange cavern where many get lost.

Complexities and chaos regarding romanticism are far apart in nature in China and the West. In the West, romanticism is considered by Karl Wilhelm Friedrich Schlegel (1772–1828), Brandes, Irving Babbitt (1865–1933), Berlin, etc. as a significant far-reaching movement in modern European history, and a "worldwide cultural phenomenon of practical relevance" (Габитова 1), though their opinions differ tremendously from each other. In contrast, the complexity of romanticism in China is caused by logical fallacy: Romanticism is castrated of its cultural intension and therefore reduced to a literary technique featuring bold imagination, grotesque plots, strange exaggeration, mythological touch, etc. Strictly, Tao's poetry and prose, except in sporadic cases, do not conform to the standards of "romanticism."

Prejudices aside, an anthropological–cultural–ideological approach to Tao reveals that he may be closer to the original spirit of romanticism. If it is acknowledged that humans across time and space share some of the appeals regarding the meta-question of "Nature and man," Tao in the medieval time and those eighteenth- or nineteenth-century romantic writers such as Rousseau, Novalis, Wordsworth, Hölderlin, and Thoreau are on a comparable basis. It is for a corrective purpose that Tao is defined here as a Nature-rooted or naturalistic romantic poet.

Liu Dajie is the first one, and the only one perhaps, who has pointed out in explicit terms that Tao Yuanming is a "naturalistic romantic poet." In his 1940s monograph *A History of Chinese Literature*, Liu writes, "As a purifier of ideologies of the Wei and Jin Dynasties, Tao demonstrates philosophically-artistically-inclined romantic naturalism" (Liu DJ 140).

In fact, the treatment of nature is a core question for the European "fundamentalist" romanticists, who, in terms of their sympathy for nature, surprisingly echo Tao Yuanming.

S. Lu, *The Ecological Era and Classical Chinese Naturalism*,
China Academic Library, DOI 10.1007/978-981-10-1784-1_3

3.1 The European Romantic Literary Tradition and Nature

"Romanticism," a key word in literary history writing, was coined in Europe as a product of modern European cultural spirit. Its origination and implications are brilliantly accounted by Isaiah Berlin, who, in his *The Roots of Romanticism*, states the following:

> The greatest single shift in the consciousness of the West that has occurred, and all the other shifts which have occurred in the course of the nineteenth and twentieth centuries appear to me in comparison less important, and at any rate deeply influenced by it. … The Romantic Movement was just such a gigantic and radical transformation, after which nothing was ever the same. (1–5)

In the late 1790s, the Enlightenment Movement, despite its glory, began to show consequences. It is out of an anti-enlightenment urge and stance that romanticism took roots. Berlin crowns the laureate of romantic forefather on Johann Georg Hamann (1730–1788), a grassroots scholar, Kant's neighbor and rival, because Hamann was "the first person to declare war upon the Enlightenment in the most open, violent and complete fashion" (Berlin 46), and he "struck the most violent blow against the Enlightenment and began the whole romantic process, the whole process of revolt against the outlook" (40). Of course, he was not alone there. Around him were many figures more visible than himself, such as Herder (1744–1803), Novalis (1772–1801), Friedrich von Schlegel (1772–1828), William Blake (1757–1827), and Rousseau.

In his attempt to define romanticism, Berlin gives an excellent account of those romantic forefathers' major ideas as follows:

Anti-scientism and anti-bureaucracy: Those romanticists opposed the abuse of science, especially the application of scientific theory in social governance, because it would lead to terrible bureaucracy. They also opposed "professionalization" that would hurt the organic community of Nature; they opposed the idea of "concepts and categories." In a word, they were hostile to all derivatives of the Enlightenment's instrumental rationality. For them, "The whole of the Enlightenment doctrine appeared to kill that which was living in human beings, appeared to offer a pale substitute for the creative energies of man, and for the whole rich world of the senses" (Berlin 43). They were of the conviction that "God was not a geometer, not a mathematician, but a poet" (Berlin 49).

Hostility to light and amity with darkness: They had such attitudes because in strong light of the Enlightenment, "When the Olympian gods become too tame, too rational and too normal, people naturally enough begin to incline towards darker, more chthonian deities" (Berlin 46). Romanticists were those who had grievances over reality and retired to their inner world; they would believe in the reverie of the depth of the poet's unconsciousness than submit to science and technology in the broad daylight.

They espoused individual spiritual freedom and urge to break away from what hinders free thought and feelings. Kant, a "restrained romanticist" in Berlin's term, believes in inborn freedom and opposed any form of rule of men over men. For him, a true genius is not to be tamed, because whoever is always dependent on others would lose his own footing and be reduced to slaves to others. Genius as Kant was, he got contempt from people like Schlegel because of the former's submission to the Duke of Weimar.

They valued spiritual freedom and human nature. Unlike Enlightenment thinkers who viewed nature as an objective neutral or hostile existence, romantics believed that man and nature should be unified as one because "human groups grew in some plant-like or animal-like fashion"(Berlin 61), because "Pulsations of spirit … are also pulsations of nature" (Berlin 99), because "Life in a work of art is analogous with … what we admire in nature" (Berlin 98), and because "the only works of art … which have any value at all … are those which are similar to nature" (Berlin 98). Hamanm was fully convinced that God made his voice heard through nature. In Hamman, "the inexpressible mysteries of nature" (Berlin 49) was mystified by romanticism.

For romantics, the natural environment, human nature, and individual spiritual freedom were damaged by the industrialization, marketization, and urbanization since the eighteenth century. As a result, the negation of social progress, "homesickness," and "the turn to nature" became eternal themes for romanticism. Berlin states:

> Romanticism is a re-awakening of the poetry of the sleepwalking Middle Ages, … an escape from the horrors of the Industrial Revolution, … a contrast of the beautiful past with the frightful and the monotonous present, … the exuberant sense of life of the natural man, … the accustomed sights and sounds of contented, simple, rural folk, … the bosom of nature, green fields, cow-bells, murmuring brooks, the infinite blue sky. (49)

The above quote indicates that Berlin believes romanticism is almost identical with salvation in terms of functionality.

A decent discussion of the European Romantic Movement cannot afford to ignore Brandes, a follower of positivism, scientism, and Europe's modernization. In his *Main Currents in Nineteenth Century Literature*, Brandes emphasizes that "romantic" was derived from the Roman language of some southern European provinces of ancient Rome, the Roman language being a mixture of many dialects and Latin, and a carrier of many ballads and sagas. Brandes's claim that "romanticism is essentially but a bold defendant of local color in literature heralded the opposition of romanticism against universal reason and law," which functioned as a cornerstone for the Enlightenment. He thinks that literary ethos and movements in the nineteenth century were, more often than not, counter-revolutionary. In his remarks on the nineteenth European romantic writers, Brandes claims that Coleridge lodged a protest against the philosophical theories of the age of the Enlightenment; that Wordsworth was too eager to negate modern civilization which ran counter to moral values, and therefore what Wordsworth, as a country man, presented in his works was stagnantly tranquil country life and mere attachment to locality; that Hölderlin, longing to escape from the artificial social structure to eternal nature, lamented the loss of bygone things, and, therefore, wrote to express his nostalgia for the lost Greece. Brandes's observation about Von Schlegel is very illuminating:

> Return to perfection is, in art, a return to the lawlessness of genius, to the stage at which the artist may do one thing, or may do another which is exactly the opposite. In life it is the retrogression of idleness, for he who is idle goes back, back to passive pleasure. In philosophy it is the return to intuitive beliefs, beliefs to which Schlegel applies the name of religion; which religion in its turn leads back to Catholicism. As far as nature and history are concerned, it is retrogression towards the conditions of the primeval Paradise. (Brandes 73–74)

Brandes's statement clearly shows his idea of "return" and "regression." Such accusations notwithstanding, Brandes has to acknowledge the achievements of European romanticism in literature and art, saying that romanticism infused vitality into almost every department of literature, and brought into the arts subject matter never before dreamed of. Ironically, those writers who, as Brandes observes, ran counter to the times, were almost unexceptionally great writers of their day. For some reason, Brandes did not explore this question: Why could those writers produce so great works though they were considered "counter-revolutionary," or, why did they, as great writers, have so drastic conflicts with their times?

Walter Benjamin (1892–1940) and Herbert Marcuse (1898–1979) carried this question forward. In a postwar context where the inner contradiction of capitalism was fully revealed, reflections on modern social progression took place. Benjamin and Marcuse were called "great romanticists" because of their repositioning of romanticism. For them, the conflict between romantic literature and modern society was caused by the incompatibility between the rules of aesthetics and those of industrial society, especially over the aesthetic position of nature. This is a core statement that explains the significant and historical value of European romanticism.

Benjamin believes that aesthetics about nature is the basis for aesthetics of arts, but the industrialized nature is hostile to man's aesthetic experience, and therefore, a form of damage to the beauty of nature, characterized by the boredom of unitary culture and the loss of the humanist historical context, has been inflicted. For Benjamin, wild nature generates the sublime, but industrialization transforms nature into a barren field of aesthetics. Benjamin's judgment was reaffirmed decades later by Rachel Carson's *Silent Spring*. In the eyes of Benjamin, scientific advancement and economic growth are not equal to the progress of society and civilization; on the contrary, such unitary "progressive force" is more like a storm from Paradise which "in the name of progress, blew human beings after the Fall farther away from Paradise, leaving behind it nothing but a wasteland of modernity" (Guo 31). As Benjamin observes, a real notion of progress should be consciousness of criticism in which nostalgia over the past, the attachment to precapitalist culture, and the "homecoming" urge derided as regression may become weaponry of the future.

Marcuse included nature in his "aesthetic dimension." He asks: Why were most of the writers and artists and especially the romantic poets of the nineteenth century critical of the ever-worsening mechanization and marketization? He points out that modern civilization treats nature as exploitable resources as it does humans, and nature is under more effective control; consequently, an artificial nature, then, becomes "the extended arm of society ... Commercialized nature, polluted nature, militarized nature cut down the life environment of man, not only in an ecological but also in a very existential sense. It blocks the erotic cathexis ... deprives man from finding himself in nature" (Marcuse 235). Romantic poets, sensitive enough, must have found the estranged nature unbearably suffocating. Their derision of and opposition to modern society was exactly a call for the nature's liberation movement. For Marcuse, "The fulfillment of man is at the same time the fulfillment, [...] of nature" (Marcuse 166).

Guy Salvatore Alitto (1942–), an American historian and sinologist, summarizes the nature of modern civilization into "the enslavement of nature by virtue of reason." As is known, no country's modernization was not based on the enslavement of nature by reason. The evaluation of a society is, to a large degree, defined by the efficiency with which nature is developed and exploited. In this case, nature has lost its glory, appeals, and position as the mother of all things. Worse still, it has become the antithesis of human beings, or simply resources to be exploited in man's favor.

Brief as Alitto's judgment is, it captures the two pillars of modern society, that is, "enslavement of nature" and "reason," one being external and the other internal. Deplorably, René Descartes (1596–1650), an advocate of rationalism, and Francis Bacon, the founder of "new instrument of science," among others, accelerated the decline of nature.

In the eyes of romantics, the Enlightenment forerunners should be held partly accountable for the destiny of nature and poetry. As regards the expectations of social progression, romanticists have faith of their own, a faith based on nature and freedom. Their conflict with modern society can be eventually attributed to their attitude toward nature.

Now it is safer to summarize the intention of the romanticism that originated in modern Europe. To put it succinctly, the modern European romantic literary movement was a trend that reacted against the ideologies and values of the Enlightenment, and that ran counter to modern industrial society. Romantic literature of this period, then, was characterized by sensibility over sense, poetic dwelling over industrial urges, spiritual freedom over material comforts, human nature over science and technology, by skepticism of progress, an urge to return to the local, to the country, to nature, to tradition, and by a touch of humaneness and sentimentality in terms of attunement realized through techniques such as imagination, exaggeration, metamorphosis, and even mystification.

3.2 Tao as an Ancient Oriental Natural Romantic Poet

In which sense can one declare that Tao Yuanming was, or is, a naturalist romantic poet?

Tao's attitude toward nature and freedom qualify Tao as a romantic. His "returning to the country life—keeping integrity in poverty—keeping to fundamental simplicity—following the True and the Natural—dwelling poetically" path and value orientation is congruent with the aspirations of European romanticism since the eighteenth century.

Undoubtedly, the oneness of Nature and man, as sought and realized by Tao, has a strong touch of romanticism. If Tao is recognized as a naturalist, he must have been at the same time a romantic. Even in terms of personality type, there exist so many striking similarities between Tao and the romantic poets of the eighteenth century.

> The truth is that they were a remarkably unworldly body of men. They were poor, they were timid, they were bookish, (and) they were awkward in society. They were easily snubbed, they had to serve as tutors to great men; they were constantly full of insult and oppression. It is clear that they were confined and contracted in their universe; they were like Schiller's bent twig, which always jumped back and hit its bender. (Berlin 131)

If one uses Berlin's description above as a frame of reference, he or she may find that the unworldly body of men may refer to Rousseau, Hamann, Novalis, Hölderlin, etc. However, one is under the impression that Tao Yuanming is also on the list. Tao served for years as secretary to a general and eventually resigned from the post of Head of Pengze Prefecture, which can be viewed as a jump back of the bent twig.

Both in conception and style, William Wordsworth (1770–1850), a nineteenth-century poet laureate, the universally acknowledged first romanticist in British literature, has, following Shakespeare and Milton, become a romanticist pattern for literary historians worldwide. Wordsworth and Tao are far apart temporally and spatially, yet they have much in common. The juxtaposition of Tao and Wordsworth has become in recent years a hot topic in comparative literature.

Tu An (1923–), a renowned Chinese poet and expert in English literature, praises Wordsworth as one who "reverently devoted his entire soul to nature and then made every effort to seek human nature in it" (Tu An 253). The low-born Cockermouth native loved nature, chose to leave the urban clamor, and settled reclusively in the lake district of Cumbria and Grasmere. Like Tao's, Wordsworth's verses are often eulogies for the idyllic landscape. Soft breezes and white clouds, the moon and the sun, dales and streams, cottages and chimneys, wild grass and flowers, and flocks and herds were part of his life. Only in nature could he have spiritual and physical freedom and soothe his anguish from the earthly world. Reminiscent of Tao's "chrysanthemum-picking" imagery is Wordsworth's "To the Daisy," which reads:

> Thou unassuming Common-place
> Of Nature, with that homely face,

And yet with something of a grace,
Which love makes for thee!

Oft on the dappled turf at ease
I sit, and play with similes,
Loose types of things through all degrees,
Thought of thy raising:
And many a fond and idle name
I give to thee, for praise or blame,
As is the humour of the game,
While I am gazing.

This presents both an Oriental "oneness of things and me" and a Western way of "denken." This has been, more often than not, the way of romanticists' lived poetic experience, which Berlin describes as "the secret and inexpressible delight of a soul playing with itself" (Berlin 16).

Like Tao, Wordsworth often captures in his poems nostalgia, attachment to the country and the delightful thought sometimes difficult rural life, and a yearning for a small peasant economy exemplified by the "Peach-blossom Springs," for he was steadfast in his belief that the country life is the most natural. Wordsworth, like Tao again, also contemplated death since death is the most intimate way of connecting oneself with nature. Tao knows this very well. Wordsworth must have also realized that death is not to be lamented because it is but a way back to Nature as the ultimate home. In the "Lucy Poems," Wordsworth frequents the topic of death, expressing the idea that Lucy lives as if she is a part of nature, unnoticed when alive, nor deplored otherwise. The return to nature is just a matter of fact. Wordsworth tended to eulogize childlike innocence as children are in his eyes closer than grown-ups to nature, therefore, are more forthright, wayward, and prone to the enlightenment by nature.

Coincidentally, the poet even compares himself to a "lone cloud":
I wandered lonely as a Cloud
That floats on high o'er Vales and Hills (Untitled, 1802)

It is a suitable simile for his mood since the drifting cloud is lonely and unchained. The cloud has the vantage point of overlooking the worldly people and their vicissitudes, and it delights people, pure and simple. In fact, the English "cloud" is not really alone there, for he is paired by Tao who writes:

All things grow up from roots on which to lie,
Except the lonely cloud high in the sky.
It vanishes as winds begin to blow,
Leaving not a trace of faintest glow.
When mist has been dispersed by rosy dawn,
The birds in flocks are flying in the morn.
The solitary bird would leave the forest last
And come back home before the day is passed.
It knows itself and keeps routes as of old,
And so, it starves and suffers from the cold.
If I can find no one to share my thought,
Why should I feel the woe and grief for naught! (141)

More desolate and determined as Tao's "lonely cloud" appears, the above-cited poem, a piece in the series of "On Poor Scholars," is similar to Wordsworth's portrayal in that they both express their righteous retreat, freedom, independence, and unrestraint, which is an expression of the "soul playing with itself."

Even the hostile reviews on them are similar in nature. Wordsworth in modern Europe and Tao in contemporary China face the same charges from radical revolutionaries: passive, feeble, backward, retrogressive, and reactionary.

They are believed to have similar literary styles. "My heart leaps up when I behold/A Rainbow in the sky." These first two lines of an untitled lyric of Wordsworth show a style, as some believe, resembling the rainbow radiating splendors from a fresh, simple, and elegant arch. As for the manner of Tao, criticisms in as early as the Song Dynasty likened it to "a purple-red cloud in the sky that folds and unfolds as it pleases" (Liu Xun, "Living as a Hermit"). The comment captures the naturalness and originality out of mundaneness and plainness.

Tu An comments globally on Wordsworth, stating:

> Wordsworth's poetry, from content to form, from plots to language, can hardly maintain itself without an attachment to, and an admiration for, nature which shelters souls and humanity. For Wordsworth, the truth of life may lie in the longing and adoration for nature. For him, nature and human life are inseparable. The life of a poet is at the same time an effort to get integrated into nature and a practice to integrate nature. (Tu An 262)

In Chap. 1, Liang Qichao uses "nature" many times in his praise of Tao Yuanming, and here in this case, Tu An treats Wordsworth in the same fashion. This points to the link between Tao and Wordsworth across time and space, as Robert Bly, a poet of the American Deep Image Group, observes: "Tao might be Wordsworth's 'spiritual ancestor'" (Li Ping 157).[1]

3.3 Tao's Foreign "Naturalistic Soulmates"

Tao Yuanming's international influence is predominantly within the Sinosphere.

The reception of Tao in Japan is the most impressive. The Japanese came to learn about Tao via *The Refined Selection of Literature* compiled by Xiao Tong the Zhaoming Prince (501–531). According to recent research in Japan, Tao's anthology was brought to Japan by Japanese students in the eighth century. Published in the year of Tenpyōshōhō (751 A.D.), *Huai Feng Zao*, or *Nostalgia of Embellishment*, a well-known collection of poems, includes Tao's "Peach-blossom Springs" and "Homeward ho!" During the dynasties of Heian and Edo, Tao's life and works became sources that informed a number of Japanese poems, and even the Mikado Saga created by drawing on Tao's lines his own "Enjoying

[1]Li (2004).

the Chrysanthemums," reading "Autumn changes while people remain the same,/ Things and people fade away./Enjoy chrysanthemums at the east fence,/and forget all the worldly intrigues." This is proof that the naturalist spirit of Tao's poetry made its way into the hearts and minds of the Japanese. Master Koubou (774–835), an idol of Japanese culture, expressed his appreciation of Tao's verse, reading "In early summer, grass and trees grow tall/With profuse foliage sheltering the hall. The flocks of birds have the fondest place to rest,/While I love my cozy hut the best." He held this as a paragon of the "aesthetics of environmental nature." After the Meiji Restoration, Miyazaki Kosyosi, a celebrated critic, classified Tao's poems in her "Going Home for a Visit" into the "Back to Nature" category, thus presenting a new pattern of literary studies. Since the twentieth century, Japanese sinologists' studies on Tao have been growing steadily.

The Korean Peninsula did not have that tradition until the early Song Dynasty (960–1127), when there came recorded intensive studies. For instance, Li Kuibao (1584–1674), a poet and philosopher, experienced similar vicissitudes in his earlier life as a well-read ambitious political career seeker just like Tao. Li was later banished due to his unbending integrity and thus retired to his retreat writing. He was abandoned to poetry, wine, and *qin* (Chinese zither) music, styling himself "Three Likes" as Tao styled him "Five Willows." Li's utter admiration for Tao can be seen from his remarks: "Ironically, my poetry, intended to follow the style of Tao, ended up a poor imitation of his. Tao's poetry is full of ease, tranquility, and soothing effect just as zither music from inside a temple in the depth of stillness." (Kuang 6) Li Zhi (1584–1674) valued Tao's poems well above others' in his gradation of Chinese poems. For him, the Han and Wei poets were the best in producing five-character poems; in terms of capturing human nature and dispositions, Tao's poetry is the best except the celebrated "Nineteen Ancient Poems." He selected and transcribed more than forty pieces and chanted them every day. Ljasin (1681–1763) viewed Tao as a great man, a sage who followed the natural running of the cosmos, saying "I read Tao's poems about seasons, and feel the nature's operation. I, as it were, was watching the river like a sage" (233). For Ljasin, "A great man is one who coalesces his virtue with that of the haven and the earth, his order with the four seasons. His individuality, though limited, is able to communicate with *qi*, for whether he opens or shuts his outlet, he does it naturally, like what Tao Yuanming wrote in his poems about seasons" (228).

However, the West does not have as strong passion for Tao. Since the twentieth century, the cultural exchange between China and the West has been predominantly unidirectional. A century's self-critique and even self-castration have almost exhausted China's traditional cultural resources while the Western cultural tidal waves have submerged almost all the Oriental fronts. What has been seen are cultural imports rather than exports. Even the poetic tradition has not fared well though China is labeled as "the kingdom of poems." The most celebrated contemporary Chinese poets such as Hai Zi and Wei An have completely ignored the long-established Chinese poetic tradition. Ironically, their list of their favorite poets and writers would include Homer, Dante, Shakespeare, Tolstoy, Goethe, Nietzsche, H. Anderson, Kafka, Whitman, Tagore, Emerson, Gibran,

Pushkin, Yesenin, Keats, Shelley, Rousseau, Bunin, Hugo, Hesse, J. Michelet, W. H. Hudson, J. Renard, Prishvin, Astafiev, and Hölderlin, of course, and a very few Chinese poets, if any. Personally, I would always be thankful for those Westerners who do sympathize with traditional Chinese culture.

In the West, France has been a fashion setter in accepting Tao Yuanming. *The Chinese Collection*, a French journal at the turn of the nineteenth century, published some of Tao's poems in French translation. A deeper bond between Tao and French writers came some time later quite coincidentally.

> Your translation of Tao's poetry is inspiring thanks to your unparalleled French competence and the simple touching beauty of these verses in their own right, whose attunement sounds so familiar to French ears! The aroma rising from our ancient lands smells just the same. (Liang ZD 288)

This quote was a reply of Romain Rolland (1866–1944) to 23-year-old Liang Zongdai, a young Chinese poet studying in France, who mailed to Rolland his 19 translated poems and several pieces of prose of Tao. Rolland confessed to Liang how excited he was about the striking similarities between the two different mentalities. Rolland's passion was shared by Paul Valery, a French symbolist, who became so enthusiastic about Tao that he wrote the preface to Liang's Les Poems de *T'ao Ts'ien*, offering his empathetic commentary on Tao's poetry as well as the Chinese cultural tradition in general: "The Chinese race is, or was, the most literarily-inclined of all"; Tao's poems present "a supreme simplicity, near to perfection." Valery's comments also kept to the keyword nature: "Examine how Tao regards nature. He integrates himself into it. He participates in it. [...] Sometimes he acts like a lover, sometimes more or less like a sage with smiles." He even takes Tao as "the Chinese La Fontaine and Virgile" (20–21).

Henri Michaux (1899–1984), author of *Knowledge for the Gouffre*; *Winds and Dust*; *The Way of Qi*; *Days of Silence*; *Towards the Serenity*; and *A Barbaric Man in Asia*, was also a fan of Tao Yuanming. Obsessed with ancient Chinese culture, Zen and Daoism in particular, the distinguished poet and writer was gentle and aloof from the world, leading a secluded and tranquil life. The very short list of his associates included a number of overseas Chinese intellectuals such as Cheng Baoyi and Zhao Wuji. With Liang Zongdai's *Les Poems de T'ao Ts'ien* in his mind when he traveled in China during the 1930s, he was deeply impressed by Tao's naturalism, a crucial nutrient for his subsequent literary creation. He confessed to Luo Dagang, a Chinese scholar visiting him, that Tao Yuanming was the most admirable ancient Chinese poet, because "Tao represents such a lofty poetic style as not to be found elsewhere" (Luo DG 7).

Thanks to the effort of Sir John F. Davis (1795–1890), a nineteenth-century Fellow of the Royal Society that Tao Yuanming was introduced to Britain. The English sinologist had a strong affection for classical Chinese literature and did extensive research on it. He translated many Tang poems and Yuan poetic dramas, compiled the Ming and Qing novels like *Haoqiuzhuan* (*A Story of Fair Ladies*), and authored a monograph on Tao Yuanming entitled *T'ao Ch'ien: His Works and Significance*, which "With an approach of psycho-aesthetics and aesthetics of reception in its analysis of Tao and his poetry, comes to a fresh conclusion" (Chen YB 3447). The twentieth century saw more English translations and studies of Tao in Britain and Australia, though not active and visible enough to be able to present Tao Yuanming effectively to the world.

Interestingly enough, the topic of Tao Yuanming emerged much later in the American academia, but it is seriously treated in textbooks. John H. Mckey's *A History of World Societies*, for one, contains a brief special biography about this extraordinary Chinese farmer, one who drank, composed poems, indulged himself in "the Peach-blossom Springs," like Thoreau of *Walden*. However, ironically, other Chinese figures *vis-a-vis* Tao in the textbook should be some "embarrassing" if significant ancient Chinese, Yang Yuhuan and Li Zicheng, were included, the ridiculousness of which exposes the way some American historians sketch the Chinese cultural tradition. In his recent publication about Chinese history, the American scholar Mark Elvin, author of *The Retreat of the Elephants: An Environmental History of China*, classifies Tao as an "environmentalist."

The limited resources on my part hardly suggest any influence of Tao Yuanming on American literature. What I do see instead is that Han Shan, or Monk Cold Mountain, an almost "lunatic" poet of the Tang Dynasty and also a naturalist poet, impressed the American readers as "the Poet of the Wild." He re-emerged as an idol for the Beat Generation.

Gary Snyder (1930–), one of the few Beat poets alive and world famous artist, is a nature lover as well as a farmer. In his interview with a Chinese magazine, he states:

> I am a laborer, and most of my nature poems are indeed related to labor. We should know about nature, about the biosphere and the atmosphere, plants and planets. One unfamiliar with nature is unfamiliar with his or her own whereabouts, like a lonely ghost … It is necessary for everyone to love nature. And it is crucial to learn something from birds or stars without necessarily being a natural historian or an astronomer. The Confucianists realized this long ago, and the first stanza of The Book of Odes demonstrates the way people feel affined to the physical world, which is thought to be negligible by many urban dwellers. (Yang 68–70)

The American poet dedicates all his creative compassion to nature, trying to offset the disorder, filth, and ignorance of modern society with the crispness, innocence, and harmony of nature. He is praised as the "poetic spokesman" of the Green Movement and is even respected as a foresighted sage dedicating his entire self to the maintenance of the ecological balance.

Snyder never confesses how his artistic career has been influenced and nourished by traditional Chinese culture. In as early as his school days, he began to read Lao Tzu, Zhuangzi, and Confucius, and came to know about classical Chinese poetry via the translation of Ezra Pound and Arthur Waley. This was the moment when he became infatuated with Monk Cold Mountain, so much so that he even followed in the Oriental monk's footsteps and spent three years in a Japanese temple. On coming back, he settled down in the northern Californian Mountains to plow, teach, and compose, living a simple, semi-agricultural, and semi-cultural life. He firmly believes that Chinese poetry will play a leading role in our common future, arguing, "Those poems created by the inhabitants of the Yangtze River and the Yellow River during the last two millennia have benefited the whole world, and will continue to enrich and enlighten us" (Yang 70).

Snyder has regrettably overlooked Tao, the most brilliant example of ancient Chinese naturalist poets, but Snyder and Tao have much in common, especially in their life choices and artistic practice. On the meta-question of humanity, they are soulmates. As a matter of fact, Tao has far many more foreign naturalistic soulmates except the European romantic writers and poets mentioned earlier.

Epicurus (241–270 B.C.), for one, is a soul mate of Tao. He was a provincial in Athens from a humble family, his father being a country language teacher, and his mother a street peddler. His contemporary scholars always derided him as being ignorant of classics and norms. Epicurus' works, however, recalcitrantly vital, has outlived those of his contemporaries partly because he, among a galaxy of Greek philosophers, followed a unique naturalistic path. Ancient Greek philosophy forked into rationalist philosophy, spearheaded by Plato, and naturalistic philosophy represented by Epicurus, who, in turn, was the icon of the most consummate state of Greek naturalistic ethics.

The Chinese academia is used to the stereotypical image of Epicurus as a hedonist, or more directly, Epicureanism. What he pursued was indeed something spiritual, a mental serenity, not unlike Tao's ideals and aspirations. The core Epicurean rationale lies in his obedience to nature: "We must not resist Nature but submit to her. We shall satisfy her if we satisfy the necessary desires and also those bodily desires that cause us no harm while sternly rejecting those that are harmful" (Epicurus 45).

Epicurus also inclined toward a simple and rustic country life. He states that "Poverty, if measured by the natural purpose of life, is great wealth," that "Extravagant wealth is of no more benefit to men and women than water is to an already full glass. Both are useless and unnecessary," that "The wise man is happy with a modest lifestyle," and that "Do not let yourself get too busy. Know your limits and do not attempt what is beyond them" (Epicurus 46, 54, 52, 50).

Epicurus might have been a good drinker. He would feel at home in Tao's hut for a drink, for according to him, the man in the midst of (his constant belching and) drunkenness believes that he is living with virtue as well.

He was too a man that knew life and death. "You are mortal," he declares, "We have been born once and there can be no second birth. For all eternity, we shall no longer be" (Epicurus 45). He remained composed about death: "When we come to the journey's end, we must be content and calm" (Epicurus 48). Before his death, he writes, as Tao did, "On this truly happy day of my life, while at the point of death, […] but against all this is the joy in my heart at the recollection of my conversations with you" (Epicurus 37). These fragmentary quotes simply resonate in spirit with Tao's philosophizing.

Tao's American naturalistic vis-a-vis are Ralph Waldo Emerson (1803–1882) and Walt Whitman (1810–1892). Emerson is to America what Confucius is to China, though the former is of a romantic type. His naturalistic aesthetics emphasizes the oneness of nature and spirit: Nature is also a being of spirit, and each natural phenomenon symbolizes that of spirit, an indication of the isomorphism of the two. He believes that all natural landscapes correspond to man's moods, which, in turn, can only be explained by those landscapes accordingly. This does not differ much from Tao's poetic motif.

John Burroughs (1837–1921), a close friend of Emerson, a nature writer and also a farmer's son, makes this appraisal: "This man had almost in excess a quality in which every current poet was lacking—I mean the faculty of being in entire sympathy with actual nature, and the objects; and shows of nature, and of rude, abysmal man; and appalling directness of utterance therefrom, at first hand, without any intermediate agency or modification" (166).

In Whitman's case, even the title *Leaves of Grass* itself can evoke permeating idyllic fragrance. Whitman believes that a happy life is always simple, and good poems are natural. Great minds think alike, indeed. Whitman also concludes that the myriad things in the universe depend on, transform, and energize each other. This is the natural law by which man should abide. A human community that turns against nature is doomed to fail. Like Tao, Whitman also showed his "anger" occasionally, as can be seen from a short poem of his in which he harshly scolded the then American president. Whitman writes:

> All you are doing and saying is to America dangled mirages,
> You have not learn'd of Nature – of the politics of Nature
> You have not learn'd the great amplitude, rectitude, impartiality,
> You have not seen that only such as they are for these States,
> And that what is less than they must sooner or later lift off from these States. (334)[2]

On the way back to nature and the original from modern society and material civilization, Paul Gauguin (1848–1903) went even farther than Tao Yuanming. Before the age of 40, the French artist resigned from his enviable position as a banker, left home, dismissed all that was tagged civilized, and went all by himself to Tahiti, a small island in the Pacific, and immersed himself in the greatness and mystery of nature. That was a place much more "primitive" than Tao's "Suli," "Donggao," and even the "Peach-blossom Springs." He nevertheless entirely integrated himself into the local tribes as a friendly neighbor, thereby culminating artistically. A confession of his is as follows:

> I am leaving in order to have peace and quiet, to be rid of the influence of civilization. I want only to do simple, very simple art, and to be able to do that, I have to immerse myself in virgin nature, see no one but savages, live their life, with no other thought in mind but to render, the way a child would, the concepts formed in my brain and to do this with the aid of nothing but the primitive means of art, the only means that are good and true. (Huret 48)[3]

When the spring of 1898 would soon see his death, he, instead of writing an elegy for himself as Tao did, completed a magnificent painting entitled "Where

[2]Whitman (2001).

[3]Huret (1891).

Do We Come From? What Are We? Where Are We Going?" By this he not only willingly accepted nature's arrangement but also offered to be a guide for all living beings as well, declaring cheerfully, "I have finished a philosophical work on this theme, comparable to the Gospel" (Huret 29). His confession and declaration are reminiscent of Tao's philosophicality about death: Once you're dead and gone, what then? Trust yourself to the mountainside.

Another "naturalist ally" of Tao's, hardly a poet though, is Carl Gustav Jung, a major contributor to psychological and archetypal criticism in the twentieth-century literary scholarship. Jung devoted his entire self to nature, particularly water, claiming that he could count on his intelligence only at waterside, which coincidentally became a footnote to the traditional Chinese wisdom that "Wise men like the waters and benevolent men mountains."

Compared with Heidegger, Jung had been influenced by Daoism in a more direct, complete, and profound way. He was of firm conviction that darkness is where brightness lurks. Jung traveled in his middle age to the "black world" of Africa, up close to those primitive tribes in Kenya and Uganda, and miraculously, had a feeling of "going back to home," thinking that it might have been his home thousands of years ago. He was told by some tribal sages that white men think with their minds and black men feel with their hearts, which greatly shocked him. David Rosen observes that "In an uncanny way, it was as if Jung were following Lao Tzu's precept: Know the white/Yet keep to the black: be a pattern for the world. If you are a pattern for the world, the [D]ao will be strong inside you, and there will be nothing you can't do" (Rosen 94).[4] It seems that Jung also followed in Tao Yuanming's footsteps in living "knowing the bright but cleaving to the dark." In his forties, Jung settled down on the picturesque upstream Zurich River, ready to replant himself there and regerminate, just like plants. His dwelling, which looked like a tower, was made of stone from Bollingen because "Jung's goal to be one with stone speaks to his life-long closeness to Nature and sense of Harmony with the Tao" (Rosen 85).

The more senile Jung grew, the greater homage he paid to Laos Tzu and Zhuangzi, so much so that he held the Daoist sages as his mentors. In his old age, Jung lived a simple life like that of the country folks, growing vegetables, cut firewood, pumped water from the well, and prepared food, as he accounts:

> I have done without electricity, and tend the fireplace and stove myself. Evenings, I light the old lamps. There is no running water, and I pump water from the well. I chop the wood and cook the food. … In Bollingen, silence surrounds me almost audibly, and I live "in modest harmony with nature." Thoughts arise to the surface which reach back into the centuries, and accordingly anticipate a remote future. Here the torment of creation is lessened; creativity and play are close together. (Rosen 128)

Even in his twilight years, Jung would often go out into the wild for a short stay. The towering snow-capped mountains blocked the vastness of the azure sky

---

[4]Rosen (1996).

and sketched a small patch of blue, leaving the wilderness even more desolate. Breathing the fresh air, Jung laments, "I think that's the last time I shall meet the mountains" (303) teardrops in his eyes. His ultimate *gnosis* (knowledge) about nature is similar to that of Tao, though, admittedly, his perception of the mountain must have been different from Tao's sight of the southern hills.

According to Rosen's account, in early June of 1961, just a few days prior to his decease, Jung had a dream that in a dark forest, the tree roots reached out from within the earth and surrounded him giving off a shining light of gold. Then and there, "Jung had become part of the Secret of the Golden Trees and firmly rooted in Mother Earth" (159). In a sense, that light might have been the glowing of Nature's mystery in Jung's soul.

In the galaxy of writers and poets, Tao's soulmates are far more than the ones listed above, and this book is not intended to be exhaustive in this regard. However, we cannot afford to overlook Rousseau and Thoreau, both of whom deserve decent discussion, because they are Tao's soulmates in almost every sense of the word.

3.4 Tao and Rousseau: From Civil Men to Natural Men

The eighteenth-century Europe saw the roaring development of industrial civilization coupled with the increasing tension between nature and man, exerting drastic impact on almost every conceivable aspect of social life, politics, and human spiritualities. How to live as humans re-emerged as a philosophical issue. In an age of thinkers, the prolific and stylistically zestful Jean-Jacques Rousseau (1712–1778), as a unique ideologist of the Enlightenment age, has produced lasting influence on the progression of human society. During his lifetime and afterward, opinions about him were and still are divided.

Unlike other scholars, Ernst Cassirer, a contemporary German philosopher, argues that all of Rousseau's works are integrated, centering around the most fundamental issue: the entanglement between "nature" and "civilization," the conflict and concordance between the "homme naturel" and the "homme artificial." What Rousseau tried to convey throughout his life was that the "homme artificial" had undermined man's natural state. Cassirer says, "The return to the simplicity and happiness of the state of nature is barred to us, but the path of freedom lies open; it can and must be taken" (54). Obviously, Cassirer's question of Rousseau is ultimately a question of "Nature and man."

Regarding Rousseau's pursuits, Christopher Kelly has a view similar to that of Cassirer. Kelly states that Rousseau's lifetime pursuit was to preserve his nature and live a natural life. In other words, Rousseau's aspiration was to a large extent about how to overcome the obstacles standing in his way back to nature, to transcend civilization-induced "denaturalization," and to realize or preserve the naturalness and wholeness of human nature.

In his "A Discourse on the Moral Effects of the Arts and Sciences," which elevated him to fame from obscurity, Rousseau questions the whole historic change to which the European Enlightenment gave rise, arguing:

> So long as government and law provide for the security and well-being of men in their common life, the arts, literature and the sciences, less despotic though perhaps more powerful, fling garlands of flowers over the chains which weigh them down. They stifle in men's breasts that sense of original liberty, for which they seem to have been born; cause them to love their own slavery, and so make of them what is called a civilized people. (Shang "SJR" 10–11)

It is a Daoist tradition to value naturalness and despise artificiality. Rousseau declares with no less finality in the opening chapter of *Émile*:

> Everything is as good as it leaves the hand, of the Author of things; everything degenerates in the hands of man. He forces one soil to nourish the products of another, one tree to bear the fruit of another. He mixes and confuses the climates, the elements, the seasons. He mutilates his dog, his horse, his slave. He turns everything upside down; he disfigures everything; he loves deformity, monsters. He wants nothing as nature made it, not even man; for him, man must be trained like a school horse; man must be fashioned in keeping with his fancy like a tree in his garden. ("E" 37)[5]

The problems such as those of "the elements" and "the climates" that Rousseau refer to here have been authenticated by mounting evidence including global warming and genetic pollution. The maltreatment of horses and trees was also observed by Zhuangzi and Gong Zizhen: The former accused those "good at training horses" of "having killed half of the horses," and the latter deplored those florists' rope binding all winter plum trees, which helped to make profit by sickening and deforming the trees ("My Plum Tree Infirmary"). No wonder Rousseau felt that, in comparison with those highly civilized French men and women, he was more like an Oriental.

Book One of *Émile* explains that the education of "natural man" is different from the education of "civil man." He portrays the latter vividly: "Civil man is born, lives, and dies in slavery. At his birth he is sewed in swaddling clothes; at his death he is nailed in a coffin. So long as he keeps his human shape, he is enchained by our institutions" (Émile 42–43). Here the "swaddling clothes" and "coffin" symbolize the bondage and damage of man's nature by the civil society that he partly builds and conforms to. The way a "civil man" suffers from civilization was keenly felt by Tao, whose "trap" and "net of the earthly world," parallel to Rousseau's "chain," and "coffin" urged him to resign, though the root reason for his retreat was that he was naturally inclined by nature and reluctant to be enslaved by institutions.

Rejecting the glory of civilization progress and safeguarding nature intact, he would rather be a lone walker groping his way in the darkness than a "civilized man" drifting around. His *Confessions* written in his later years seem to be repentance for stepping wrongly into civil society and becoming a "civil man" as well

[5]Rousseau (1979).

as reflections on how he came to a rude awakening, and pulled himself back to Nature to a natural man. For Cassirer, *Confessions* are practically a record of a civil man who, frustrated everywhere in age of reformation, made every effort to return to Nature as a natural man and to seek his origin in an excruciating way. Kelly finds that "Jean-Jacques Rousseau's project in his *Confessions* is to transform the way his readers look at the world by offering a picture of an exemplary human life" (Preface).

Rousseau, out of his own life experiences, believes that human beings are born good, and that "primitive men" must have been more simple, more innocent, more humane, and therefore, freer and happier. Meanwhile, he is not oblivious of the reality that modern people decidedly cannot return to natural primitivism. What he seeks is the way modern civil men would change their nature so that their conscience could be reclaimed, and that his state and society could be reorganized in accordance with natural laws. Rousseau's efforts ended up soliciting a battery of backfire. He was not only banned and banished by the government, but also denounced by the church and rejected by the public who even slung stones at him. On his way back to Nature, Rousseau, exhausted, keenly felt the failure of social integrity. What he has achieved is his personal return, in his lonely senility, to his natural integrity and wholeness.

About his probe into the process that transforms civil men back to natural men, Rousseau wrote millions of words. His reflections on "civilization" and "nature" cover the state system, social formation, political philosophy, educational policies, legal systems, and religious ideology, contributing to the "Enlightenment" by taking an "anti-enlightenment" stance. His writings are far more comprehensive than Tao's just over a hundred pieces of poems and prose. However, when it comes to their intent to flee the worldly cage and become a "natural man," Tao seems to be more determined and straightforward.

Though most literary historians tend to think Tao was not fairly judged in their day, yet, Xiao Tong's remarks, among others, are the most in-depth and incisive. In his *Biography of Tao Yuanming*, Xiao Tong offers an interesting episode: Tan Daoji, a high-ranking official of the royal court, went to visit Tao. "Why self-deprivation at a time of affluence," Tan asked, "And why not give up farming and go back into your political career?" Tao declined point-frankly, saying, "It's the least of my aspirations." Tao threw away the pork, a token of respect, that Tan had gave him. That was a gesture of his self-exile from this-worldly society. In "Preface to *The Anthology of Tao Yuanming*," Xiao Tong modestly acknowledges the possibility of a happy life made possible by a civilized world, saying, "It is fine to enjoy creature comforts, dance and music, convenient chariots, luxurious clothing and accessories, though worries may come along." He profusely eulogizes the natural life characterized by resignation to Nature's transformation. For him, Tao's life choice was due to his sagacity and philosophicality. Tan pointed out that some glorious figures involved in political entanglements such as Su Qin and Gongsun Yang had not made the world safe and secure, but they themselves were executed; in stark contrast, Tao retreated to Nature and the pastoral, keeping to the way, taking delight in farming, and feeling no shame for impoverishment, which was conducive to social climate and the purification of the hearts, so much as that the corrupt would become clean and cowards brave.

Surprisingly, Rousseau and Tao, confronted with the same meta-question of "Nature and man," have the same aspiration to returning to natural men from civilized men. As a result, they have some interesting similarities regarding a series of specific questions. For instance, they both oscillated for a while between continuing their political engagements and divorcing themselves from politics; they both chose to become "hermits"; they both, in their prime years, harbored the ambition to cure social ills and benefit the community. When Tao was 40, he resigned from the post of Head of Pengze Prefecture, and, at the age of 44, Rousseau discarded his political career because he realized the corruption of the privileged and the incompatibility between personal talent and political order. Tao refused to "make curtsies for the salary of five Bushels of rice," and Rousseau refused the offer of an annuity from Louis XV, who had wanted to win over men of letters. Rousseau declares, "After that how could I dare to speak of independence and disinterestedness?" ("CC" 319)

According to Zhu Guangqian, what is described in "The Peach-blossom Springs" is similar to Rousseau's "natural state" (243). Rousseau, in his late years, bid adieu to his circle of associates in Paris, returned to the wild for a life in a hermitage, since, as he himself confessed, he had caged in an environment not agreeable to him for 15 years. April 9, 1796 witnessed his departure from the city for good, and "this could reasonably promise me a happy and durable life in the one my inclination had made me choose" (Rousseau "CC" 338). Rousseau is firmly convinced that "Cities are the abyss of the human species. At the end of a few generations the races perish or degenerate. They must be renewed, and it is always the country which provides for this renewal" ("Emile" 59). In his retreat offered by Madame D'Epinay, Rousseau writes the following:

> Although it was cold and there was still some snow, the earth was beginning to vegetate; one saw violets and Primroses, the buds of the greens were beginning to sprout, and the very night of my arrival was marked by the first song of the nightingale, which made itself heard almost at my window in a grove of trees that touched the house. Since I had forgotten my transplantation, upon awakening after a light sleep I still believed myself to be in rue de Grenelle, when this warbling suddenly made me shiver and I cried out in my rapture, "At last all my wishes have been accomplished." ("CC" 333–334)

In fact, what Rousseau expresses is very similar to Tao's confession in "Homeward ho!": He decided to abide by the law of nature to lead a life of nature. So, both Rousseau and Tao have returned to the idyllic, or Nature, or the origin of life, or the dwelling of the souls. Nevertheless, Rousseau's and Tao's return are also different, though the differences do not negate the likeness in their wisdom and conscience demonstrated in their treatment of the relationship between nature and man.

In Tao's day, when farming in China had grown into maturity, one was bound to be "baptized" at birth by farming civilization. However, in eighteenth-century France and Switzerland where Rousseau lived, the Enlightenment Movement had taken roots, the foundation for industrial civilization had been already laid, urbanization was under way, and the French capitalist revolution for power was about to burst, Rousseau, as an intellectual concerned about world affairs, could not live in a hermitage as Tao did. What Rousseau could do was to wander between Geneva, Paris, and Venice even though he had intense longings for nature.

In eighteenth-century France, democracy almost took shape. In that social milieu, Louis XV would not be bold enough to have a dissident beheaded at the slightest of offences. Intellectuals did have some freedom of speech. The French were fairly tolerant to Rousseau's "fooling around," including his masturbation, his voyeurism, his visiting prostitutes, his love triangle, his abnormal sexual relationship with Madame Warren, though, he, from the viewpoint of a natural man, had an excuse for all this. Similarly, Tao, in his youth, also suffered from bodily desires. In his "Prose on Idle Emotions," Tao expresses his unquenchable longing for his beloved female body, saying that he wishes he were a pair of silk shoes so that he would touch a woman's feet, or, alternatively, a mattress so that he would carry and caress her soft body for long. Obviously, as natural men, both Rousseau and Tao had strong bodily desires, but Tao, in his day, violated the social norms. For this reason, his "Prose on Idle Emotions" was not included in Xiao Tong's anthology, and, worse still, constantly denounced by later critics.

Rousseau and Tao have different personalities: The former is more sentimental, sensitive, vulnerable, introvert, and indecisive; the latter more composed, serene, and detached. Rousseau spent almost his entire life swaying between seclusion and society, for he was unable to become completely independent of those nobles; consequently, he was often fooled, insulted, and plagued by anxiety, indignation, fear, and depression. In contrast, Tao maintained his composure because he resigned to Nature's transformation. Only at the last stage of his life did Rousseau lose his hope about the civilized society: The king was not reliable, nor the marshals, nor the ladies of nobility, not to mention those sly politicians. He keenly felt that the most human-hearted affectionate man was banished from society through a "consensus." Lonely as Rousseau "in exile" was, he enjoyed the freedom he had never dreamed of before. He exclaims, "Fifteen years of experience have instructed me at my own expense; back now under the sole laws of nature, I have regained my original health through them" ("RSW" 95). He describes his regained life as follows:

> All alone, I went deep along the winding paths up the mountain, and, passing from wood to wood and rock to rock, I finally reached a refuge that was so hidden that it was wilder than anything I have ever seen in my life. Black fir trees mixed in and all entwined with huge beech trees, several of which had fallen over with age, formed an impenetrable barrier around this refuge; all that could be seen through the new gaps in this dark wall was sheer rock faces and terrifying chasms which I only dared look into while lying flat on my stomach. From the mountain crevices could be heard the cries of the horned owl, the little owl, and the barn owl; at the same time, a few rare but familiar little birds lightened the horror of this solitude. I found there seven-leaved toothwort, cyclamen, nidus avis, the large laserpitium, and a few other plants that delighted and amused me for a long time; but gradually overcome by the strong impression made on me by the things around me, I forgot about botany and mosses, and began dreaming more freely, imagining that I was in a refuge unknown to the whole universe where my persecutors would never be able to unearth me. ("RSW"114)

Again, Rousseau's account is similar to that of Tao in his "Homeward ho!" Rousseau, in his twilight years, rejoiced for his delivery "from all earthly passions that the tumult of social life engenders" ("RSW" 494), and for his rediscovery of all the enchantments of nature, thus realizing his final return. What he differed from Tao is that he was still considering himself as an exile in a refuge.

The stability and ease, which Rousseau in his old age tried to seek, are akin to Tao's "tranquility" and "abandonment." Rousseau believes that all natural men idle as civilized men toil. He resorts to idleness as a way out of civilized corruption and misery. He defines idleness this way: "The idleness I love is not that of a do-nothing … It is both that of a child who is ceaselessly in motion while doing nothing and, at the same time, that of a dotard who strays when his arms are at rest" ("CC" 537). He always wanted "to wander nonchalantly in the woods and in the country, here and there to take up mechanically, sometimes a flower, sometimes a branch; to graze on my fodder almost at random" ("CC" 537). Undoubtedly, this bears some likeness to Tao's chrysanthemum picking.

Different from laziness, idleness as such is a way of life that is still based on self-reliance. The difference between Tao's and Rousseau's self-sufficiency is that the former was engaged in farming, delightfully close to nature though painstaking, and the latter, born and bred in the city, chose to transcribe sheet music (Rousseau was even the inventor of numbered musical notation.). It is said that Rousseau had transcribed more than ten thousand pages of sheet music within the seven years prior to his death.

Though their writing styles are different (Rousseau had a multitude of political writings, and Tao mainly poems), yet their likeness seems more significant, the likeness being their relation to Nature as mentioned earlier. Cassirer states that at a time when "verse had become a mere shell to which thought conformed. … all living sources of poetry dried up" (85), Rousseau alone lifted "the spell that rested on French language and poetry was broken only by Rousseau. Without creating a single piece of what might properly be called lyrical poetry, he discovered and resurrected the world of lyricism" (Cassirer 85). André Maurois (1885–1967) claims that "He [Rousseau] was the first who introduced green things into our literature" (Sainte-Beuve 163). The reason beneath Maurois's claim is that, as Cassirer observes, "In himself, this feeling had grown out of the direct communion with nature which he had cultivated from the first awakening of his spiritual self-awareness. He taught nature to speak once more, and he never forgot her language, which he had learned in his childhood and adolescence" (Cassirer 85). For Cassirer, Rousseau's writing is "[…] as if enveloped in the atmosphere of pure sensitivity to nature. Here man no longer simply stands 'over against' nature—nature is not a drama which he enjoys as a mere spectator and observer; he dips into its inner life and vibrates with its own rhythms" (86). The very underpinning for Rousseau's writing style, in the eyes of Cassirer, is nature, so much so that "This style does not yield or submit to the strict standards which French classicism had set up as the fundamental laws of the art de penser and the art de'ecrire. It persistently slips away from the strict line of argumentation" (127) These remarks on Rousseau's writing style remind us of some ancient critics' remarks on Tao's style, most typical of which are "uniquely self-sufficient without bothering to conform to established norms" (Huang Tingjian), "spontaneously overflowing from within the heart without studied conception" (Zhu Xi), "a presentation of himself in natural writing" (Xu Xueyi), "naturalness in chanting"(Yuan Xie), and "a colorful cloud changing freely as it pleases" (Ao Taosun).

Rousseau's and Tao's attitudes toward "death" are also on a comparable basis. For Tao Yuanming, to live is to conform to Nature, and to die is just a different form of conformation to Nature. Rousseau in his old age also came to this realization. Rousseau says that one, at its birth, begins to move toward life's end. For him, "If there is any study still appropriate for an old man, it is solely to learn to die" ("RSW" 30). Rousseau wanted to adjust himself in the rest of his life to a state that he would maintain at death. After reflections on earlier theories about death, he realized that death is just a natural process, so for natural men, there is no fear of death, which is close to a natural state of being. For Rousseau, the anguish of death and fear about the impending death are just imaginations of civilized men, or results induced by the education of medical science. About death, what he felt is "less sadness than a peaceful languor which even had its sweetness" (Kelly 155). Such a state may be described by "a mixture of sadness and joy," which Hongyi, an enlightened Chinese Buddhist monk, felt just before his death.

Tao wrote himself significantly elegant elegies in an autumn and died in the early winter. Rousseau did not write himself any elegy, but he experienced a "rehearsal" of death. In October, 1776, the news about Rousseau's sudden death was spread in Paris, which excited his rivals. Voltaire, for one, cursed him by declaring that "[h]e ate like a devil and caught an indigestion; he died like a dog" (Friedlander 123). Having known much about the world and the civilized men, Rousseau was not the least annoyed by this, saying calmly, "Now I am here, in all tranquility at the bottom of the abyss, a poor unfortunate mortal, but impassive like God himself" ("RSW" 480). Such composure about death can only be found in mortals such as Tao and Rousseau, who, knowing what life and death mean, are willing to be one with Nature.

Being one with Nature, a natural man may "live" forever. Is it not true that the "quasi-natural" Rousseau outlived the "highly civilized" Voltaire? Is Tao's longevity in the sense of his being present in Chinese culture not due to his identity with Nature?

Presumably, Rousseau knew nothing about Tao Yuanming or Tao's writing, but nobody can deny that they are soulmates across time and space on the question of nature and man.

Shang Jie describes the elderly Rousseau as follows:

> On a small country road, the senile Rousseau, neatly clad, waist bent, a walking stick in his right hand, and a bouquet of wild flowers (chrysanthemums perhaps) in his left hand, was gazing at the horizon, fearless, intoxicated. (Ye & Wang 334)

If the above description is changed into a real physical portrait, who can say that it is not a Tao Yuanming of the eighteenth-century France?

3.5 Tao and Thoreau: The Creation of Dreamy Nature and Freedom in a Poetic Ambience

Unfortunately, social progression was not willed by people such as Tao Yuanming and Rousseau. Over the past two centuries since Rousseau's death, the conflict between man and nature has become more intense. People in the East and in the West seem to be unexceptionally immersed in their triumph over nature, which has led to the near-collapse of the ecosystem of the earth. In short, nature is jeopardized.

"The death of nature," a coinage in the 1960s, has been constantly reaffirmed, if not confirmed. Representative works in this regard include Carolyn Merchant's *The Death of Nature* in the late 1970s and Bill McKibben's *The End of Nature*. A review of the history of human civilization reveals such a course of development: After humans departed from the original primitivism and simplicity, the tension between Nature and man became ever-increasingly intense until man's overwhelming triumph over nature in the industrial age. Paradoxically, this triumph is at the same time of man's fiasco, because they seem to forget, if not choose to overlook, that man and nature once were one. In this sense, the end or death of nature means the vanishing of human nature shaped by nature out there; moreover, it also means the disappearance of the value of life.

Max Scheler (1874–1928) claims that as biological beings, humans are the *cul-de-sac* for nature. For him, humans, who have been evolving for millions of years, seem to be so naive: They, in a rapidly developing industrial society, have cornered Nature and themselves into an impasse. However, facing the impending death of nature, Scheler reminds people that there is still a ray of hope because man, who is of dualism, is "both an impasse and a way out simultaneously" ("SWS" 1376). Fortunately, some people, whether they are of the East or the West, have been trying to look for a way out of this impasse.

In the mid-twentieth century, when there was an even worse plight of nature, the appeal to the salvation of nature and man developed into an ever-growing ecological movement. The prophet of this movement was Henry David Thoreau (1817–1862), a guardian of the naturalness of human nature.

As a naturalist poet and ecological writer, Thoreau has become an idol for many Chinese poets and writers. In 1949, *Walden* in Chinese translation was published in China, but readers paid little attention. After its republication of the revised edition in 1982, it became an immediate sensation. Over the past decade, there have appeared more than ten versions, and predictably, many more versions are coming. *Walden* and what it represents have proved to be extremely enchanting to the Chinese readership.

In the late twentieth century, Hai Zi (1964–1989) and Wei An (1960–1998), two shining Chinese poets, among others, were great worshippers of Thoreau. Inspired by Thoreau's valuation of Nature, Hai Zi wrote a poem entitled "Thoreau Has Brains" as a tribute to Thoreau; and Wei An worshipped Thoreau so much that he would not write without referring to Thoreau. In his "After the Sunrise," he mentions Thoreau dozens of times, saying that his spiritual joy and tremor created by *Walden* would not have been aroused by other works because "Walden teaches me to live simply and to resist the temptations' of money worship. It helped me to establish a faith and a way of life for the rest of my life" (Wei 117). He also confessed that, natural amity being between Thoreau and himself, he was converted to Thoreau.

It is hard to understand why some of the younger generation poets and writers such as Hai Zi and Wei An do not admire Tao Yuanming more than Thoreau. Is Tao not a better, earlier, and closer example in simplifying life and despising money? In the late twentieth century, hardly any poet in China paid attention to Tao. In contrast, in Europe and America, immediately after the commencement of industrialization, pro-Nature voices began to resound like those of Wordsworth, Coleridge, Blake, Rousseau, Schlegel, Novalis, Emerson, Thoreau, Whitman, etc. out of their conscience. In colonial India, there emerged Tigor. Unfortunately, in modern China, such voices could hardly be heard.

At present, "the death of nature" is already an occurring reality in China. "Save-nature" voices of the poets are still weak. It comes as no surprise that Tao Yuanming has experienced a second death, following his real death about 1600 years ago, in the hearts of the younger generation poets.

What is thought-provoking is that when the younger generation Chinese poets are paying homage to *Walden*, Thoreau is expressing his longing for Chinese philosophy and sages as can be seen from his quotes from Chinese classics once in a while, whether for argument or ornament. He cites Confucius, Mencius, Zengtse, *The Golden Mean*, *The Great Learning*, and even inscriptions on a bronze wash basin owned by a sovereign of the Shang Dynasty (17th–11th century B.C.). Not all these quotes are to the point, but Thoreau's inclusiveness should be appreciated. The pity, if any, is that the Chinese classics Thoreau cites in *Walden* are unexceptionally Confucian, not necessarily because of Thoreau's personal preferences, though. In his day, Thoreau only had access to Joshua Mashman's *Confucius*, David Collie's *The Chinese Classical Work Commonly Called the Four Books*, and Pauthier's *Confucius et Mencius*, which are all Confucian classics. Considering Thoreau's naturalism, his *Walden* might have become different if he had exposure to the works of Lao Tzu and Zhuangzi. If Tao and Thoreau had got to know each other, they would become pure-minded soulmates, pulling at each other's strings. At the International Symposium "Transcending Thoreau: the Literary Response to Nature" held in Beijing in October 2008, I gave a keynote speech before some scholars from America, Italy, India, Poland, France, etc., emphasizing that Tao Yuanming and Thoreau be juxtaposed from an eco-critical perspective. My key points are as follows:

Over the past 60 solid years since Thoreau's introduction into China, *Walden*, as an example of literature's closeness to Nature, has influenced the Chinese readership profoundly. Thoreau, widely admired in China, was held as a prophetic sage of the ecological age. Literature's closeness to nature has been an established tradition in China. Abhorring established social institutions and longing for the oneness of Nature and man, Tao Yuanming, a great Chinese poet of about 1600 years ago, chose to go back to a simple farming life after his timely retreat from his political career and composed, when his farming allowed, many significantly beautiful poems. Like Thoreau, he epitomizes a way of life and a notion of life that is conducive to ecological harmony.

Earth having entered into the "Anthropocene," the ecological plight now is one hundred or even one thousand times more severe than in Thoreau's or Tao Yuanming's day. From the West to the East, the much-talked-about crisis of literature or the end of literature is essentially connected with the crisis of ecology or the end of nature. As literature is the study of man, it should also be ecology that deals with Nature and man.

An ever-increasingly obvious fact is that human activities are accountable for the ever-worsening ecological crisis, and that man's human value orientation will define how the ecology will evolve. Therefore, Tao's and Thoreau's naturalistic spirits should be explored and exalted, and literature should get involved in the campaign to save the earth and, as a result of that involvement, be saved. All those in the literary undertaking should be united in their commitment to maintaining and nourishing the ecology in a literary way.

A born ecological cosmopolitan perhaps, Thoreau wished to collect all people's ecological wisdom to buttress man's combat with the worsening ecological crisis as he stated that it is necessary to offer a Bible of humanity by combining the Chinese, Indian, Persian, Hebrew, and other holy scriptures and writings. Thoreau's unfinished undertaking should be carried forward. Chinese scholars are expected to draw on Western experiences on the one hand and to explore their own national ecological legacy for Thoreau's "Bible of humanity" on the other.[6]

Presumably touched by my speech, a Western female scholar asked me to write down on her notebook the Chinese character name of Tao Yuanming. Scott Slovic (1960–), a renowned American eco-critic, maintains that "the local" should be treated as the substantial core of green thinking, saying, in Thoreau's words, "Direct your eye inward … and be expert in home-cosmography" (Slovic 183).[7] He passed a message to me from the other end of the globe that he hopes to read the eco-critical monographs on Tao Yuanming as soon as possible.

It is not unnecessary to pay homage to Thoreau the green sage before we move on.

Thoreau sold a boat at 7 dollars to N. Hawthorne (1804–1864) and taught him how to row it. Hawthorne has a vivid description of Thoreau as follows:

> He is a singular character – a young man with much of wild original nature still remaining in him… He is as ugly as sin, long-nosed, queer-mouthed, and with uncouth and rustic, though courteous manners, corresponding very well with such an exterior. But his

[6]Based on my keynote speech.

[7]Slovic (2010).

> ugliness is of an honest and agreeable fashion, and becomes him much better than beauty, for two or three years back, he has repudiated all regular modes of getting a living, and seems inclined to lead a sort of Indian life among civilized men. (Sayre 1155)

In 1847, Thoreau, 30 years of age, described his life for the members of his Harvard class this way: "I am a School master—a Private Tutor, a Surveyor—a Gardener, a Farmer—a Painter, I mean a House Painter, a Carpenter, a Mason, a Day-Laborer, a Pencil-Maker, a Glass-paper Maker, a Writer, and sometimes a Poetaster (Sayre 158)."

Emerson's observation about Thoreau is also interestingly accurate. In his eyes, Thoreau, disappointed by an age of industrialization, commercialization, and urbanization, chose to run counter to what is perceived as social progress. Emerson remarks, "Instead of engineering for all America, he (Thoreau) was the captain of a huckleberry party" (Sayre 1159).

Despite the racial, temporal, and geographical differences, Thoreau and Tao have much in common, especially on the meta-question of "Nature and man." Some superficial similarities (reading, plowing, weeding, picking flowers, cloud-watching, listening to birds' warbling, etc.) aside, there exist at least five deep-level coincidences between them.

(1) The rejection of social institutions, detachment form society, and skepticism about civilization progress

Intoxicated in Nature, they, with poetic romanticism, both distanced themselves from the social institutions of their day and retired to their dreamy world of freedom and nature. Tao's refusal to bend for royal court salary is mentioned earlier. Thoreau, versatile and promising, refused all offers of decent jobs as perceived by society and became an "idler" in Concord. He resisted, at least spiritually, many aspects of the rising American industrialization, including the railway, the bank, the post, and the press, which were considered as sure signs of progress.

Thoreau lamented that what he wanted were flowers instead of steel ingots from factories. Tao confessed that he was like a bird in nature but was caged by society. If social institutions are always expressions of the civilization of an age, Tao's and Thoreau's anti-institutional practice is a rethinking and criticism of that civilization. Undoubtedly, such rethinking is indispensable for the health of social institutions.

(2) Returning to farming, and looking for meaning and underpinning for life in nature, Tao's return to the idyllic has been accounted earlier. Compared with Tao's decades of farming, Thoreau's farming life in the woods by Walden over the course of only 26 months was more like an experiment or superb live art. Nevertheless, Thoreau's experience with nature, within and without, is as profound as his emotions are true.

Justifiably, in an age of rising industrialization, it was much more difficult for Thoreau to return to farming. Difficulties notwithstanding, Thoreau, at the age of 28, went into woods by Walden with a borrowed axe, built himself a dwelling, opened up clearings for farming, thus beginning his poetically naturalistic experiment. He wanted to live a simple natural life like nature itself. He grew beans, cut wood, plucked wild fruits, and went fishing sometimes for minimum creature comforts.

Thoreau uprooted his life from the industrialized and commercialized world and returned to more primitive farming pattern, because he believes primitive agriculture is more closely connected to Nature; therefore, it has some degree of holiness. Thoreau writes, "By avarice and selfishness, and a groveling habit, from which none of us is free, of regarding the soil as property, or the means of acquiring property chiefly, the landscape is deformed, husbandry is degraded with us, and the farmer leads the meanest of lives" (156). Thoreau wishes that man should have more time to communicate with nature and, in doing so, acquire the highest of spiritual pleasures. He gives a poetic account of one of his idle tranquil beautiful days as follows:

> Sometimes, in a summer morning, having taken my accustomed bath, I sat in my sunny doorway from sunrise till noon, wrapped in a reverie, amidst the pines and hickories and sumaches, in undisturbed solitude and stillness, while the birds sang around or flitted noiseless through the house, until by the sun falling in at my west window, or the noise of some traveler's wagon on the distant highway, I was reminded of the lapse of time. I grew in those seasons like corn in the night, and they were far better than any work of the hands would have been. They were not time subtracted from my life, but so much over and above my usual allowance. I realized what the Orientals mean by contemplation and the forsaking of works. (Thoreau 106)

Thoreau's life as such explains through personal experience the Daoist idea that "Push far enough towards the Void. Hold fast enough to Quietness" (Lao Tzu 33), which enables one to live to Nature. In fact, this idea is also expressed in Tao's writings. Tao writes in his poetic line, "The view provides some veritable truth, but my defining words seem to me uncouth" (Tao 113) Similarly, Thoreau writes, "The day advanced as if to light some work of mine; it was morning, and lo, now it is evening, and nothing memorable is accomplished. Instead of singing like the birds, I silently smiled at my incessant good fortune" (Thoreau 92).

For Thoreau and Tao, to integrate themselves into nature is the most meaningful and beautiful activities.

(3) Safeguarding freedom and the purity of the soul by poverty

After Tao returned to farming, his life became more and more difficult to the degree that he was not even decently clad and fed, but he, impoverished, took delight in tilling and maintained his integrity. He kept to his way because of his realization that poverty and simplicity were the gateway to enlightenment, freedom, and Nature.

Thoreau's deliberate poverty, though in an experimental way, helped him develop profound conceptions about poverty. For him, Concord dwellers' wooing fortune was to live in "others' brass." Undoubtedly, their goals were achieved, more often than not, at the sacrifice of their independence and freedom of life, just as Thoreau observes, "Most of the luxuries, and many of the so called comforts of life, are not only not indispensable, but positive hindrances' to the elevation of mankind" (Thoreau 90).

Thoreau and Tao had their own pleasures. For instance, Tao might spend a day this way: "When I lie in leisure beside the northern window in May or June, I would feel like a man living in [the] ancient times" (Tao 273). That was free though rare enjoyment for Tao. In *Walden*, Thoreau emphasizes several times the equality between rich and poor in terms of the enjoyment that Nature brings, saying, "The setting sun is reflected from the windows of the alms-house as brightly as from the rich man's abode; the snow melts before its door as early in the spring" (Thoreau 306). Thoreau finds that the poorer are more likely to live an unrestrained life. In fact, it is for that unrestrained life that Tao chose the poor farming life. Thoreau might be able to sympathize with Tao's choice, if he had known about him, as can be seen from his observation that "The ancient philosophers, Chinese, Hindu, Persian, and Greek, were a class than which none has been poorer in outward riches, none so rich in inward" (Thoreau 12).

(4) Valuation of spiritual freedom, substitution of inward riches for outward possessions, and transcendence over reality for higher pursuits

Regarding the meta-question of Nature and man, both Thoreau and Tao try to seek spiritual freedom by replenishing the outward want with inward richness. For Tao, this finds expression in his "remoteness of the heart," his nonchalance about rises and falls in life, and his resignation to nature's transformation. Tao's unrestraint in the Daoist sense of the word and his philosophicality are expressed in this writings in the form of romantic transcendence over reality, the Peach-blossom Springs being a good case in point.

For Thoreau, spiritual freedom is expressed in his self-esteem and solitude. Like Tao's detachment from society, solitude, which is good for health, is what he wants, as he confesses, "Society is commonly too cheap. … I find it wholesome to be alone the greater part of the time. To be in company, even with the best, is soon wearisome and dissipating. I love to be alone. I never found the companion that was as companionable as solitude" (Thoreau 128). Thoreau draws an analogy between himself and things out there to highlight his sweet solitude, saying:

> The sun is alone. …. God is alone…. I am no more lonely than a single mullein or dandelion in a pasture, or a bean leaf, or sorrel, or a horse-fly, or a bumblebee. I am no more lonely than the Mill Brook, or a weathercock, or the north star, or the south wind, or an April shower, or a January thaw, or the first spider in a new house. (Thoreau 167)

By immersing himself in solitude, Thoreau achieves spiritual freedom and beyond. Average men find loneliness, especially prolonged loneliness, hardly bearable, simply because they demand too much. In contrast, sages fear no loneliness because of their enriched inner world. To enhance this point, Thoreau cites a quotation of Confucius that "From an army of three divisions one can take away its general, and put it in disorder; from the man the most abject or vulgar one cannot take away his thought" (Thoreau 307). As for the inward–outward relationship, Thoreau says that one can sell his clothes but not his thought. That is to say, clothes, which are outward, are not indispensable for spiritual freedom, but thought is. In his diaries, Thoreau confesses that he has not done much to improve society; however, he wishes to cultivate pearls in shells. His work *Walden* is the very crystallization of his rich inner world, or an air castle built with poetic language. At the end of *Walden*, Thoreau writes:

> I learned this, at least, by my experiment; that if one advances confidently in the direction of his dreams, and endeavors to live the life which he has imagined, he will meet with a success unexpected in common hours. ... If you have built castles in the air, your work need not be lost; that is where they should be. Now put the foundations under them. (302–303)

Undoubtedly, Thoreau's air castle and Tao's Peach-blossom Springs, similar in nature, are just dreamy poetic dwellings they created, symbolizing their aspiration to the integrity of Nature and human society. This means that they, by resorting to their artistic imagination, have realized poetic transcendence over disappointing social realities.

(5) Contributions to their respective national literature

In addition to beans of various kinds and elevated tastes, pursuits, and ideals, Tao and Thoreau have a good literary harvest, especially poems and prose. What they present and represent in their writings is out of nature and dedicated to nature. For some reason, Tao inclines toward the idyllic, whereas Thoreau prefers the wild. As a result, the former has initiated the Chinese *tianyuan* poetic tradition, and the latter has set the fashion of "nature prose" in American literary history. In short, their literary contributions cannot be exaggerated too much.

These five aspects are the major similarities between Thoreau and Tao. Besides, another interesting connection can be established between them, too. Thoreau writes, "If the fairest features of the landscape [Walden] are to be named after men, let them be the noblest and worthiest men alone" (185). But, the fact is, as Thoreau continues, "[O]f all the characters I have known, perhaps Walden wears best, and best preserves its purity. Many men have been likened to it, but few deserve that honor" (181). I wish I could ask Thoreau personally whether Tao, if he had known about Tao at all, deserves that honor.

Undoubtedly, there exist some differences between Tao and Thoreau. The former is a great drinker, a man of family burden, and predominantly a poet; the latter is a teetotaler, a lifetime single man, and predominantly an essayist.

Tao tried to shun the rigidly hierarchical order of his day, the harm to man's true nature inflicted by those hypocritical canonized classics, and the endless sufferings caused by incessant wars. Therefore, his return, having a touch of helplessness, seems to be a deliberate hiding of the sage's glory. In contrast, Thoreau's retreat in the woods is not an escape of any kind. He returned in order to gather evidence and experiential knowledge to challenge the modern society.

Due to some distractions, Thoreau could not always completely immerse himself in nature as Tao does. Consequently, his writings, expressing some anger, anxiety, and protest in some cases, make him look like an occasional "angry young man." For this reason, Thoreau's image of simplicity has also a touch of the glory of a "world savior" as opposed to Tao's self-healing of the heart.

Their paths leading to the oneness with nature differ tremendously. Thoreau, labeled as "transcendentalist" though, tends to resort to reason and knowledge on his way to discovering the great truth of nature, whereas Tao realizes this via intuition and contemplation in a completely unconscious way. To put it differently, Tao's unintendedness and effortlessness makes Tao Yuanming Tao Yuanming, whereas Thoreau's endeavor makes Thoreau Thoreau.

Anyway, the way they deal with Nature, among other things, has elevated them to the status of sages. Unfortunately, Nature, to which they are so attached, is almost past salvation. Hopefully, this is not the real end of Nature.

Chapter 4
The Evolving Perception of Nature and the Death of Tao Yuanming

Admittedly, the Chinese literary history did not begin with Tao, but Tao's nourishment and shaping of Chinese literature can only be matched by that of Qu Yuan (352–281 B.C.), Li Bai (701–762), and Du Fu (712–770). His naturally inclined personality, his natural philosophy, and his romantic naturalism are unique in Chinese literature.

The social pattern may be determined by man's attitude toward Nature.

When ancient Chinese thinkers advocated "the oneness of Heaven and man," the confrontation between Nature and man had already emerged. But this confrontation was not encouraged, and no "conquering nature" theories, if any, were established as state policies or ideology. Instead, the ancient Chinese tried to bridge the rift between Nature and man by establishing a series of codes of conducts, norms, and aesthetic yardsticks, thus maintaining the continuity of the agricultural society for several millennia. By today's standards, this is called "backwardness". In contrast, in the West, where prevails the idea of "the conquer of nature," people tend to exploit Nature by bringing reason into full play, thus creating an affluent industrial society just over a short span of three hundred years. This is called "progress". However, as a piece of Daoist wisdom goes, "In Tao, the only motion is returning." (Lao Tzu 87) Huamn's linear progress has entered an "impasse". It may help to turn to China's 5000-year-long agricultural civilization, for in doing so, we may find an alternative way out of this impasse.

Tao is usually considered as a *tianyuan* poet and the founder of the *tianyuan* poetic tradition in China. This is not wrong, but, in a broader sense, Tao can be viewed as the poetic symbol of China's farming-and-reading tradition. This tradition runs from the time of *The Book of Poetry* through the entire Chinese literary history. Informed by *The Book of Poetry*, Tao's poetic tradition finds expression in the poems of the Tang and Song dynasties, especially in themes such as "being a woodcutter in the wild," "farming", "homesickness", "integrity in poverty," and on "simplicity and naturalness" as an aesthetic pursuit. His tradition is still active in contemporary Chinese literature in general, and local color or regional literature in particular.

S. Lu, *The Ecological Era and Classical Chinese Naturalism*, China Academic Library, DOI 10.1007/978-981-10-1784-1_4

It is a thought-provoking question that Tao, seemingly out of sight or within reach, has become "the Other" in today's industrial society.

Indisputably, an agricultural society is usually more closely connected to Nature than an industrial society. An agricultural society like China in the past five millennia has much wisdom about, and attachment to, Nature. The perception of Tao as the avatar of Nature, which is accurate, is exactly based on the "Nature and man" relationship. The literary spirit that Tao represents evolves with the status of Nature in the history of Chinese civilization in general, and in the history of Chinese literature in particular.

4.1 The Evolving Perception of "Nature" in the History of Chinese Literature

It is of special value to put side by side the writing of history over the past one hundred years with the writing of the history of Chinese literature of the same period. Since 1904, when Lin Chuanjia and Huang Ren published their *A History of Chinese Literature*, great achievements have been made in the writing of the history of Chinese literature, but there is also much room to be desired. Historical studies across the world have made a significant turn to pluralism, as indicated by the criticism of Eurocentrism and the attention paid to underprivileged civilizations. Unfortunately, Western yardsticks are still used to measure the history of China. Such a practice shows that history studies in China are rarely based on the traditional Chinese cultural spirit, thus lacking autonomy in pluralism though having universality. The mainstream Chinese scholarship still sticks to the old-fashioned conception of history, and, worse still, rejects the new conceptions of history that have emerged since the middle of the twentieth century. One of the expressions of such practice in the writing of the Chinese literary history is the ignorance of Nature or Nature dimensions. As discussed earlier, "Nature" has a holy status in traditional Chinese culture. In a sense, classical Chinese philosophy is natural philosophy, which is different from the Western tradition. Such ignorance, from an eco-critical point of view, should be corrected. However, we have to admit that it is an unforgivable mistake in terms of the preservation of China's literary legacy.

Earlier on, histories of Chinese literature usually contained contents of "Nature", though it is presented each in a different way. For instance, Li Li's *A Review of Chinese Literature*, published in 1928, has chapters titled "Literature and the Environment" and "Local Colors of Literature". Yesasia's *The History of Chinese Literature*, published in 1930, includes two Nature-related chapters, namely "Poets and Mother Nature" and "Literature and Climate". Xu Xiaotian's *The History of Chinese Literature* contains contents such as "The Nature School," "Geographical Elements in Literature," and "Geographical Distribution of Men of Letters in China." Later on, the writing of the Chinese literary history grew more mature, but "Nature" either disappeared or was reduced to, if at all, subject matter, the setting, or even a passive object to be conquered. For instance, on a single page in the 1982 edition of *The Evolution of Chinese Literature* compiled by Liu Dajie, there should be up to 30 words bearing the meaning of "combat", "control", or "conquer".

In most Chinese literary history writings, labor is believed to be the origin of literature. However, labor means man's struggle against Nature. Mythology, presumably the archetype of literature, usually reflects man's strong will to combat nature. Conveyed by many Chinese myths such as "Nv Wa patching up the broken sky," "the Jingwei bird filling up the sea," "Houyi shooting down the extra suns," and "Kuafu chasing after the sun" is man's power against Nature. These myths appear frequently in literary history works of various kinds. However, myths expressing the amity between Nature and man such as "Pan'gu separating the Earth from the Sky," "the Death of Huntun (Chaos)," "Liezi Riding the Wind," "Cangjie's invention of words," "the Red Dragon giving birth to the Sagely Yao (a legendary emperor of antiquity)," and "the Mysterious Bird giving birth to Shang" are usually excluded from these works. Even those myths that are included are often interpreted too randomly. According to Ye Shuxian (1954), the sun-chasing Kuafu was not a martyr who sacrificed in his struggle against the elements; he was trying to reconcile the *yin* and the *yang*, which symbolizes the incessant motion of the Way (Ye 139).

In fact, in Chinese literature, Daoist aesthetics is far more influential than the thought of Xunzi (313–238 B.C.), a realistic Confucianist, and Wang Chong (27–97), a materialist philosopher. Nevertheless, in most Chinese literary history works, the content about Lao Tzu and Zhuangzi accounts for only half of that about Xun Zi and Wang Chong. Besides, due to their idea of the oneness of Heaven and man, Lao Tzu and Zhuangzi are denounced as being passive, pessimistic, superstitious, and reactionary, but Xun Zi and Wang Chong, for their ideas that there is division between Heaven and man and that man will triumph over nature, are praised as fearless atheist warriors and great thinkers. Wang Chong, in particular, exerted little influence on Chinese literature originally; his "Lun Heng", which elevated him to fame, is not without contradictions and logical fallacies. Ironically, after he was praised as a great materialist thinker by Hu Shi (1891–1962) during the May Fourth Movement period, Wang Chong has become what he is today. Those Western scholars who have yet to have a better understanding of Wang Chong even compare him to the Chinese Voltaire or, more boldly, Plato. In contrast, Dong Zhongshu (176–104 B.C.), who espouses the idea of the oneness of Heaven and man, the interconnectedness of everything, and the abeyance of Heaven to have Dao established, and who influenced the literature of the Han dynasties profoundly, is either ignored or criticized by contemporary literary historians. Gong Pengcheng laments this situation, declaring, "It is mistaken to trace the origin of literary criticism of the Six dynasties to Wang Chong" (Gong 19). The reason why Wang Chong finds favor in today's literary historians is that his Lun Heng clearly expresses the idea of the division of Heaven and man and the idea of pragmatism, which, in turn, conforms to the mainstream ideas since the Renaissance.

The writing of "mature" works of Chinese literary history is based on the identify with the development pattern of the West defined by the division of nature and man as opposing entities and by productivity and efficiency with which nature is exploited. In this sense, literary development means how it "diverges" from nature. Consequently, there emerges in the literary history writing a disparity between the intrinsic value of literature and prejudices of social consciousness. Simply put, in modern times, "The writing of the Chinese literary history coincides with China's effort to establish its place in the new international framework" (Dai 2).

However, the Chinese were once exposed to the world in a wrong way and at a wrong time. In the late Qing Dynasty, when China's closed door was forced open by the Opium War, the Chinese, whose Way and tradition were declining due to the raging political corruption, found all of a sudden that they were facing a powerful arrogant West. When the Chinese intellectuals began to look up at the West, most of their self-confidence disappeared. In such a context, where China's admiration for, and imitation of, Western civilization had to be preceded by the derision and rejection of China itself, the reception of the Western civilization had to be realized by the uprooting of Chinese civilization. During the May Fourth Movement period, many of the forerunners of the Chinese democratic revolution, including Hu Shi, Lu Xun, Chen Duxiu, and Qian Xuantong, to name but a few, were involved in this "self-uprooting" campaign. Subsequently, imported into China were some modern Western ideas such as, visibly, "science", "democracy", "dialectical materialism" "historical materialism", and, less visibly, "pragmatism", "utilitarianism", "scientism", "nationalism", "dualism", and "social determinism".

During the May Fourth Movement period again, the mainstream of the Chinese literary history writing was "scientific-positivistic" school spearheaded by Hu Shi. What are beneath his call for "vernacular literature" and "popular literature" are science, democracy, and social Darwinism. For Hu shi, since the development of tools determines social progress, the vernacular Chinese language is more advanced classical Chinese, and therefore, the replacement of classical literature by vernacular literature represents literary progress. By this standard, the ancient poets and essayists, obviously writing in classical Chinese, now become literary sinners, and, at the same time, the novels and dramas of the Yuan, Ming, and Qing dynasties, which emerged partly due to the flourishing urban commerce, especially the handicrafts industry, represent the culmination of Chinese literature. Worse still, many long-established literary traditions have been disintegrated or deconstructed by bold hypotheses; case studies are crowned as "positivism". Ironically, in the communist China, Hu Shi is criticized in political campaigns, but his theory of literary history has been completely accepted, including his exaltation of Wang Chong. Traces of Hu's influence can be identified even in Mao Zedong's thought on literature and arts, especially the idea of "popularization" and "easy and lovable forms of expression."

After the founding of the People's Republic of China in 1949, the writing of literary history shifted from the principle of "positivism" to that of "class struggle" and its practice, thus minimizing the scholar's independent thinking and personal style. Therefore, *Shuihu Zhuan*, better known as *All Men Are Brothers*, were derided as a how-to book encouraging peasant uprisings; *Hong Lou Meng*, or *The Story of the Stone*, became a scroll of portraits describing class struggle; and even the divination of texts in *I Ching* (*The Book of Change*) became the testimony to the oppression of slaves by their owners. In such a climate, the evaluation of a writer was, first of all, based on his class origin and political stance plus his or her contribution, if at all, to "social progress". For instance, the merits of Tao Yuanming were believed to be his engagement in physical labor and his refusal to "wallow in the mire" with the ruling class; his demerits were believed to be his resignation to Nature and his unwillingness to transform society. A Tao Yuanming as such is vastly different than the Tao Yuanming as the avatar of Nature.

If the writing of a nation's literary history has to be rooted in the nation's traditional cultural soil and spirit, including cosmology, mythology, theory of being, value orientation, and aesthetics, China has the least reason to overlook Nature dimensions. Unfortunately, Nature has almost lost its legitimacy in the writing of China's classical literature. However, there have also emerged several "dissidents" in terms of literary history writing, which is a welcome sign.

Lin Geng (1910–2006), a Chinese poet, is an early dissident in this regard. He published his *A History of Chinese Literature* in May, 1947. An overflow of his true feelings and independent thinking, it is a book that completely discards the academic stereotypes of his day. It argues that Chinese literature is characterized by femininity, the natural, inclination toward the *tianyuan*, harmony, and the golden mean, which is associated with agriculture and the pictographic Chinese characters. In his treatment of the relationship between Nature and man and that between literature and Nature, Lin, thanks to his poet personality, often breaks away from established norms and therefore has discovered the real significance and subtleties of Chinese literature. For instance, in his discussion of *The Book of Poetry*, he argues that "The gentle man's rites and music are affectionate; the 'wild' man's rites and music are natural. The acculturation of the 'wild' man is to be appreciative of healthy pleasures and to know that life is the end, not the means. That is how the rites of the Zhou (Confucius's time) developed" (Lin 45). Lin highly praises Tao Yuanming, saying that Tao embodies the culmination of the Wei and Jin ethos and that "Tao marks the maturity of classical Chinese poetry because he, as the first poet to appreciate Nature so profoundly, responds to Nature's health and therefore lifts himself out of suffering." (Lin 126) For Lin, it is due to Tao's influence that classical Chinese natural philosophy is perfectly expressed. He believes that "The aspiration to the wild Nature and to the simplest revelations of Creation marks a different phase of Chinese literature and arts" (Lin 126).

Another dissident is Wolfgang Kubin (1945–), a renowned German poet, critic, and sinologist, who, according to himself, "has devoted over the past four decades his entire self to Chinese literature" (Kubin Preface 1). In his *The Empty Mountain: The Evolution of the Notion of Nature in Chinese Literature* in 1985, he states that the notion of Nature in Chinese literature appeared much earlier than in European literature, but Chinese scholars' studies of this remain very inactive. He notes that the notion of Nature in Chinese literature experienced three stages, namely the pre-Qin and Han dynasties, the Six dynasties, and the Tang Dynasty, after which this notion has seen no development due to the rise of urban literature such as novels and dramas. Kubin has a difficult time comparing the notion of Nature in Chinese literature with that in Western literature. He, from a Western point of view, equals Nature in the Chinese sense of the word to "scenery", "landscape", "the environment", or "the subject of the field of labor," thus deconstructing the rich intension of Nature in the Chinese tradition, though, admittedly, such a practice makes his argumentation more easily manageable. Even in primitive society, "Nature" is not a purely physical environment, but, rather, an entity that already has features of a spiritual being. In ancient Chinese literature, it is not true that the spiritualized, aestheticized nature is preceded by the real material nature. Kubin confesses, "My preferences and refusals only represent myself. … It is not intended to be universal nor everlasting." (2).

The last dissident worth discussing in this case is Hu Lancheng (1906–1981), an overseas cultural wanderer. His *A History of Chinese Literature*, first published in 1977, is a work solidly based on traditional Chinese culture, especially classical natural philosophy.

Hu believes that the way of literature is to follow the Way of Nature. With this as the point of departure, he has established five basic principles for the writing of the Chinese literary history, namely (1) Nature has will and soul; (2) Nature has changes of the *yin* and the *yang*; (3) Nature is the unity of the finite and infinity; (4) Nature is the unity of necessity and contingency; and (5) The development of Nature is circular, not linear. (Hu LC 3–4) For Hu, if the law of Nature is also the law for literature, then the height a nation's literature and culture can reach is determined by its attitude toward Nature. He claims that real literature should be "plain and honest in its treatment of Nature" (Hu LC 20). In his eyes, only literature, and Chinese literature alone, can reach out into the mysteries of Nature which science and religion cannot.

On the relationship between Nature and man, Hu does not think highly of the West, claiming that unlike the European Moon as captured in Beethoven's "Moonlight Sonata" or the American Moon on which Apollo landed, the Chinese Moon, as depicted in Tang poetry, is the most beautiful. Nor does he think highly of the modern times. He believes that there is no sure literary progress. For him, Nature is God; being close to Nature is being close to God, and therefore ancient literature, which is close to Nature, is the best. For him, *The Book of Poetry*, in particular, is of the highest grade because it expresses the sublimity of Heaven, the Earth, and man. In contrast, in modern times, due to the over-development of science and technology, "the taste of food is that of monosodium glutamate, colors are those of chemicals. Who knows the true sounds, the true tastes, and the true colors of Nature? Even children are deprived of their faculty of perceiving simple and natural beauty. Modern people are cut off from Nature, or God" (Hu 75). Therefore, poetic and literary spirit in modern times is on the decline.

To be fair, despite Hu's personal integrity and political stains, his work, sometimes superficial, impulsive, high-sounding, less organized, and as affectionate as an old hippy's performance, is lively, and an excellent rare case in terms of the writing of literary history.

At a time when the nature exploitation-driven pattern of social development is challenged, should the writing of literary history be changed accordingly? It is time that Nature dimensions, once ignored, should be evoked and revitalized. This inevitably entails a rewriting of literary history.

A latest judgment by the international scholarship is that the earth has entered into "the Human Period." Unlike the Cambrian Period, the Devonian Period, the Jurassic Period, and the Cretaceous Period, all of which are purely geological terms, "the Human Period" is no longer a purely geological term; instead, it covers all aspects of the connection between human society and Nature, including economy, politics, security, education, culture, faith, and what not. Literary phenomena and literary history also need rethinking from a global perspective, because, in so doing, humans may be able to discover a way to help alleviate, if not eradicate, the ever-increasing tension between Nature and man.

4.2 The Origin and Development of Tao's Poetic Conception

Epitomizing the Zeitgeist of their times, great men of letters are unexceptionally rooted in their national culture. In Tao's day, the dominant ethos is the metaphysics of the Wei and Jin.

Within the Chinese scholarship, opinions on the metaphysics of the Wei and Jin are divided. The most typical negative evaluation is that "metaphysics ruined the country," implying that the social chaos was simply caused by a group of intellectuals. Obviously, this represents the ruler's attitude. In China, there is always "the rule of power," but never "the rule by philosophy." Confucius tried in vain to deliver his idealistic social governance by virtue of his philosophy. Similarly, Lao Tzu just served on a lowly position corresponding to that of the library curator. Knowing much about the vanity of the earthly world, Zhuangzi lived in seclusion writing allegories. Many intellectuals of the Wei and Jin, including some high-ranking officials, were not serious about being officials, nor were they concerned about their official responsibilities; what they did were playing the Chinese zither, playing chess, drinking, and exchanging ideas on Heaven, the earth, gods, and man, on tai chi, and qi, on life and death, on being and non-being. As history shows, those powerful rulers of the Wei and Jin failed to make the country a better place; their time is the darkest of ages in history of China. Unexpectedly, the men of letters and philosophers at that time brought forth the most profound and subtle philosophy and offered examples of the most humane and individualistic ways of life as well as the most refined, beautiful, and creative literature and arts, including Wang Bi's philosophy, Qi Kang's music, Gu Kaizhi's painting, Wang Xizhi's calligraphy, Xi He's theory on painting, Zhong Rong's, Lu Ji's and Liu Xie's critical theory, and, of course, Tao Yuanming's poetry. It is virtually a Renaissance of the Chinese medieval times; it is almost another case of "the flouring of one hundred schools of thought" that the pre-Qin era once saw. This situation is closely related to the rise of metaphysics.

Zeitgeist of the Wei and Jin, metaphysics is rooted in the legacy of I Ching, Lao Tzu, and Zhuangzi, which combined, are the very fountainhead of classical Chinese natural philosophy. Later, Buddhism, after its introduction into China, began to experience localization as it was filtered and transformed by Daoism, thus having Chan Buddhism in China we know today, which contains part of the spirit of classical Chinese natural philosophy. The metaphysics of the Wei and Jin did not reject Confucianism; instead, it incorporated and transformed Confucianism and Daoism, "Nature" being the focus. Centered around "Nature" was a hot topic: social institutions and Nature, or man and Nature.

Tang Yongtong (1893–1964) categorizes the metaphysical philosophers of the Wei and Jin into three types according to their differentiated attitudes toward Nature, despite their shared valuation of Nature. The first type is the "moderationist school" represented by He Yan and Wang Bi, who argue that social institutions come out of Nature. For them, since being comes out of non-being, Nature is the root and rites branches; in other words, Nature is an absolute ultimate being, but rites are just a means that conforms to the law of Nature. The second type is the "radical school" spearheaded by Ji Kang and Ruan Ji of the Seven Sages of the Bamboo Grove. They declare that institutions should be transcended and Nature followed because, for them, institutions do harm to human nature. They refused to cooperate with the royal court in any form, and, as a result, were ruthlessly oppressed. This was a lesson later metaphysical philosophers such as Xiang Xiu and Guo Xiang learned the hard way, so they tried to reconcile the conflict between social institutions and Nature, arguing that institutions are also a form of Nature. As a matter of fact, this idea offers much expediency in manipulating Nature. This is called the "moderate than moderationist" school by Tang Yongtong. (Tang 107–108) Strictly speaking, this "moderate than moderationists" ended up compromising, if not surrendering.

After a comparison of Lu Ji's *Wen Fu*, or *On Literature* and Wang Bi's *A Primer to Lao Tzu*, Tang Yongtong finds that though Lu Ji deals with literature, whereas Wang Bi discusses Heaven and the earth, "their rationale and approaches are basically the same" (Tang 188). Based on his reading of the opening chapter of Liu Xie's *Wenxin Diaolong*, or *The Literary Mind and the Carving of Dragons*, Tang points out that "Writings are not identical with Nature itself, but they are coeval with Heaven and the earth because 'language originated in Tai Chi, the primal beginning' and 'word patters are the mind of Heaven and the earth.'" (Tang 189) Tang also points out that regarding the value of literature, there has been the division between "literature as the carriers of the Way" and "literature as the overflow of feelings," the former being instrumental or pragmatic as represented by Cao Pi (187–226) and Han Yu (768–824), who believe that writings should serve social governance, and the latter being aesthetic in the sense that "Through writing, one can perceive the value of life and the universe, enjoy Nature, … and express the unity of man and Nature" (Tang 189). For literature of the latter category, it does not intend to convey the Way, but the Way is contained and conveyed unconsciously.

The way of Nature, which prevails in the metaphysics and literary theory of the Wei and Jin dynasties, is at the core of the ethos of the times. What is Tao's attitude toward Nature? This question, already discussed in earlier chapters though, is worth further elaborations.

Since the start of the new era, marked by the implementation of the reform and open-up policies in 1978, Luo Zongqiang's and Yuan Xingpei's studies of Tao Yuanming have remained the most significantly corrective and incisive.

Luo thinks that Tao's contribution includes, in terms of subject matter, his introduction of the idyllic life into poetry and his initiation of the style of "ease and aloofness". Before Tao, men of letters also wrote about Nature, expressing their longing for hermitage, "but they only had their bodies in nature, not their hearts because they just enjoyed Nature and had their minds soothed by Nature without becoming one with Nature" (Luo "HLT" 166). Unlike his predecessors, Tao lived in Nature and became one with it, since "Nature interacts with his soul and is unified with his life. The beauty of landscape is expressed in the blurring of man and Nature" (Luo "HLT" 166). As for Tao's "ease and aloofness", it means "the natural overflow of the blurring between man and Nature" (Luo "HLT" 169), or the naturalness of his writings. The two aspects of Tao's contribution can be summarized as writing about Nature with naturalness, thus realizing a real return to Nature.

Laudably, as early as the early 1990s, Yuan Xingpei's studies of Tao are focused on Tao's philosophizing, stating that "Tao is a poet and philosopher, and, because of this, is superior to most poets" (Yuan 1).

Undoubtedly, the connection between Tao's philosophy and the metaphysics of the Wei and Jin has to be explored if one intends to study Tao's philosophizing. Influenced by Chen Yinque and Rong Zhaozu (1897–1994), Yuan Xingpei adopts "naturalism" and "neo-naturalism" to describe Tao's philosophy (2). "Nature" is the focus of Yuan's contemplation in his work. For him, "The core of Tao Yuanming's thought is the worship of Nature" (Yuan 49). The expressions of this core include "his application of the worship of Nature in every conceivable aspect of his personal and social life, his use of Nature to gauge reality and to offset the hypocrisy of institutions" (55). One of the unrivaled beauties of Tao's poetry lies in "its truthfulness or naturalness" (61); therefore, "Both Tao's poetry, philosophy and life have reached a realm of Nature in the evaluative sense of the word. He is eminent among ancient Chinese poets and thinkers" (56). Yuan argues that Tao's valuation of nature is part of the legacy of Lao Tzu and Zhuangzi and that Tao inclines toward Ji Kang's and Ruan Ji's "discarding social institutions and following Nature." This is an accurate observation. It is worth pointing out that unlike Ji Kang's and Ruan Ji's radicalism, Tao's philosophy is rooted the moderate notion of "non-being", thus absorbing the finest of the metaphysics of his day to the maximum without being touched by the affectation and fickleness found in many of his contemporaries.

Miraculously, the lowly, far-from-being-prolific Tao Yuanming represents the culmination of the ethos of his day. Chen Yinque praises his literature and thought as "first-rate of all time." Similarly, Fung Yu-lan, a philosopher himself, believes that "Tao is of the highest gradation among all Jin celebrities and literati. … His poetry conveys the best and the highest of metaphysics as he represents the best of unrestraint. … He reached out into a realm free from sadness and joy" (Fung "V" 316).

Previously, the Chinese scholarship tended to emphasize that the Wei and Jin dynasties was an age of the awakening of self-consciousness, of the discovery of personal life, and of the construction of subjective personality without realizing that it is also an age of the reawakening of man's Nature-consciousness, of the rediscovery of Nature, and of man's conscious reimmersion into Nature. Zong Baihua (1897–1986), a poet and aesthetician, realized this inadequacy as early as the 1940s. He notes that "People of Jin discovered Nature outward and their own affections inward. … They discovered the beauty of Nature and of personality" (Zong 186). Tao is one of the discoverers listed by Zong. It is due to Tao's emergence that Chinese literature began to have a beautiful chord of Nature, man, and literature.

Ultimately, Tao's status in literary history is determined to a large degree by the Nature thought he represents, and Nature's significance in literary history is predominantly epitomized by Tao's poetic conception and personality. Thus, Tao is a hallmark in the evolution of China's literary history writing.

Regarding Tao's influence, many believe that he was always overlooked in his day based on such observations: He is not mentioned in Liu Xie's *The Literary Mind and the Carving of Dragons*, nor in "the literary circle" prescribed by *The History of Jin*. However, subject to stereotypes, socializing, discourse power, publicity, channels of communication, etc., the visibility of a writer within a particular period is not proof to his or her real value. It was no easy job for Tao to rise to some fame in his day considering that he was just a retired lowly official, an impoverished farmer, a hermit, and a self-exile poet, and that he had no intimidating social status like that of his contemporary poets such as Yan Yanzhi and Xie Lingyun, nor did he have any literary circles to cling to like "the Seven Men of Jian'an" or "the Seven Sages of the Bamboo Grove," nor did he make any sensation as Lu Ji, Lu Yun, Guo Pu, and Jikang did by either showing their "irregularities" of various kinds or presenting to the world their defiance. The ultimate yardstick for a writer's value is the potentials contained in, and realized by, his or her works after death. The truth is that Tao Yuanming was already viewed highly in his day, let alone in subsequent dynasties. His contemporary Yan Yanzhi remarks that Tao's "Homeward ho" expresses his abandonment and aloofness and his true aspiration. This remark is close to Tao's core of spirit. Shen Yue (441–513) gives an account of Tao's daily routine, including "playing unstringed zither," "filtering wine with a scarf," and "reposing by the north window," important resources for studies of Tao even today. While Xiao Tong's biography of Tao and his preface to Tao's anthology themselves are proof enough to the Prince's admiration of the commoner poet. Xiao Tong's praise for Tao can be seen from the following remarks:

> Erudite, and simple in writing, Tao has rare high tastes; therefore, he is eminent and self-contented in his pursuit of the true. … He is steadfast in his pursuit, kept to his own Way, and feels no shame for his farming and impoverishment. Who can match Tao's sageliness and integrity? … I love Tao's poetry and prose so much so that I can hardly go without

> having them at hand. Thinking of his great virtue, I wish I had lived in his day and would have collected more of his pieces. … I said that readers of Tao's writings may rid themselves of their earthly ambitions, lowly desires, impulse to embezzle public fund, and timidity. Tao is more than an example of following benevolence and righteousness and of discarding one's political careers. (Liu ZW 20)

Bao Zhao (415–466) and Jiang Yan (444–505) argue that there existed "Tao's style" that could be modeled. "During the Northern and Southern dynasties, Tao's influence was ubiquitous. … Many men of letters were acquainted with Tao's writings and experiences and, therefore, referred to Tao very often in their own writings." (Liu ZW 20)

What is the origin of Tao's poetry? In his *Tastes of Poetry*, the observation by Zhong Rong (468–518) is the most controversially influential: Tao's poetic conception was influenced by Ying Qu (190–252) and his style by Zuo Si (250–305) who was known for his powerfulness and sublimity in his writing.

In tracing the origin of Tao's poetic conception, one is not expected to be confined to only a particular fountainhead poet; one should place the origin of Tao's poetic conception in a broader cultural context and larger literary tradition. In this case, the origin in question should be the "nature" thought or spirit found in "the Ninetieth Ancient Poems," *The Book of Poetry, Lao Tzu, Zhuangzi, The Confucian Analects, The Book of Rites, Shang Shu, The Book of Change*, etc. Invisible, that thought or spirit always hovers over the sky of China, akin to "immortals" in Hu Lancheng's term.

Chinese literature in general is attached to Nature because of the long-established agricultural civilization. In such a civilization, Nature is also human's body and human's spirit and soul, in turn, are also the soul of Nature; therefore, man builds themselves a dwelling between Heaven and the earth, that is, "the *tianyuan*", whose earliest builders include the legendary Chinese ancestors such as *Youchao* the inventor of housing, *fuxi* the inventor of hunting and fishing, *Suiren* the inventor of fire, *Shennong* the agriculturist, and *Nv Wa* the sky-fixer. When earthlings, fatigued and frustrated, find that they have no way to go to the paradise in Heaven though they wish to, they may return to "the *tianyuan*" as both a physical dwelling and a spiritual anchor. Chantings about the *tianyuan* abound in Chinese literature, especially in classical poetry. For instance, *The Prayers of Yiqi*, one of the earliest Chinese poems, reads: "Let the soil go to its mound, let the water go to the valley, let pests harm no crops, and let vegetation grow where they belong." It is about the building of the dwelling. Another piece reads: "I rise as the sun rises; I nestle down as the sun sets. I dig a well for water; I till the land for food. Why should I need a king?" Besides, in the poetry before the Wei, homesickness or nostalgia over the homeland is a very common theme. This theme is rooted in the idyllic country life complex, which, as a primitive image and yearning, lies in the individual's subconsciousness and the Chinese nation's collective unconsciousness. If there has to be a specific origin for Tao's poetic conception, it should be the nation's complex about "a country life of Arcadian simplicity and contentment." This is the "archetype" for Tao's poetic imagery and conception.

For Jung, an "archetype" or a "primitive image", as a race's shared mindset, can be the result of evolution of certain animal instincts or the trace of temporally remote social activities in the collective mind of the race. It is the "crystallization and concentration of repeated psychological experiences" (Feng C 61). The Chinese nation's complex about "a country life of Arcadian simplicity and contentment" was as active in primitive agricultural society as it was in Tao's day as the *tianyuan* poetry before Tao and by Tao indicates.

According to Jung, "A nation's archetype or the primitive image awakens in the most creative personalities; it reveals itself in the illusions of artists, in the inspirations of thinkers, and in the internal experiences of mystics" (64). Jung believes that such extra-personal unconsciousness is akin to the omnipresent and omniscient "soul" in Greek philosophy. Tao was not a mystical person, but he definitely was a most creative artist and thinker; as a result, the complex and image of "the country life of Arcadian simplicity and contentment" is vivid and clear in him and in his poetry. This can be explained by an extreme piece of wisdom of Jung's: the possession by ghosts occurs when the specter of collective unconsciousness possesses an individual body, thus making him a poet. For him, it is not that Goethe made Faust but that the specter of Faust as Germans' psychological archetype made Goethe. In a similar way, we can say that Tao has become what he is perceived to be is because he is possessed by the specter-like complex of "a country life of Arcadian simplicity and contentment" and the unquenchable urge to return such complex generates. In this sense, the specter of Tao is a part of the soul of the Chinese nation.

After Tao, the imagery of "returning to Nature" and of "returning to the country" grew dramatically in Chinese literature.

"Nature" as a theme of literature can be found in ancient mythology. When human beings just came out of the jungle, they were still closely tied to Nature like other mammals, though humans already had hands that could use material tools and had language that encouraged the growth of their spiritual world. As their spiritual world internalized the external world, they still could not clearly distinguish gods from human beings, or the external world from their inner world. For them, Heaven, the Earth, gods, and man exist in the same way in the same organic holistic world, which is, in fact, chaos.

In ancient Chinese mythology, "Chaos" is a God, muddleheaded, powerful, and senior in the nation's genealogy tree, whose posterity includes Zhurong the Fire God, Gonggong the Water God, Houtu the Earth God as well as the gods of cloud, of birds, and of wind. The images of deities at that time usually have more properties of Nature like those of fauna and flora. For instance, the Emperor of Yan, or the legendary ancestor of the Chinese nation, is the Sun God with a cow's head.

As human civilization evolved, the blurring of Nature and man gradually disappeared because man realized that they are a relatively independent being vis-a-vis Nature. Thus, Nature became the "environment". However, in an agricultural society like ancient China, man, still rooted in the earth, had strong spiritual reverence for and emotional attachment to Nature. In social activities, what man consumed was mainly the surplus of the earth's ecological system. The "chaos" in the childhood of humanity gradually became an ideal perfection, informing man's spiritual Utopian designs. It is around this time that Nature became a theme of artistic representations.

Fu-bi-xing was, and, to a large degree, still is, the basic thought pattern and techniques applied in Chinese writings. It is proof enough to the harmony between Nature and man in literature. It is widely accepted that *fu* means the direct representation of things, events, and feelings (reminiscent of realism), *bi* refers to the representation of events via that of things (akin to impressionism), and *xing* is the expression of thought and feelings by the perception of things (a touch of romanticism). Things, in this case, mainly refer to natural existences and occurrences such as fauna and flora, the elements of nature, celestial bodies; events refer to personal and social activities and happenings; and feelings refer to all human affections. To better explain this, it is not unnecessary to quote the first stanza of the opening poem of *The Book of Poetry*, reading:

By riverside are cooing
A pair of turtledoves;
A good young man is wooing
A fair maiden he loves.[1]

The turtledoves' billing and cooing, which is a thing, and the lad and the lass' sweet nothings, which is an event, create a beautiful aesthetic experience, which is a feeling.

Qu Yuan writes in the same vein, unconsciously perhaps, in his *Li Sao*: "Th' winds whistling and leaves falling make a rueful sight! Oh, would I be yearning for your grace but in vain?" (Zhuo 55). In fact, the techniques of *bi* and *xing* are even better employed in *The Verse of Chu* than in *The Book of Poetry*. To some degree, the view of Nature as a subject and not as an opposing object does not sever man's tie to nature, including their spiritual reverence and emotional attachment.

In the Wei, Jin, and the Six dynasties, Chinese literature saw its second period of blooming. Informed by the philosophical debates with metaphysics as the core, more aesthetic experiences about Nature became possible in literature, and *tianyuan* poetry and landscape poetry emerged as a distinct poetic genre, the hallmark of literature of that period. The zeitgeist of the times is best captured and conveyed in poems of this nature, highlighting the unique status of Nature in man's social and spiritual life. For instance, Tao's poetry and prose depict, among other things, the desirable harmony and unity between Nature and man. For this reason, over the past centuries, it has remained a norm to admire, eulogize, imitate, integrate, inherit, and dream of Tao as represented by Meng Haoran (689–740), Li Bai (701–762), Bai Juri (772–846), Zheng Gu (851–910), Wang Anshi (1021–1086), Su Dongpo (1037–1101), Huang Tingjian (1045–1105), Wang Ruoxu (1174–1243), to name but a few, though there were sporadic anti-Tao voices.

[1]Xu Yuanchong's translation is adopted here. Tr.

In short, because of its Nature dimensions and its rootedness in "the country of Arcadian simplicity and contentment," Tao's poetic conception will never die.

4.3 A Turning Point in Tao's Reception

During the Ming and Qing dynasties, there were mounting records about Tao Yuanming, but his reception was not as encouraging as in the Tang and Song dynasties. Besides, the reasons for his reception became more complex. Such changes came along with the re-evaluation of the relationship between Nature and man and the value of Nature itself. They find expression in the Chinese aesthetic experience and artistic life. Certainly, Tao, rooted in naturalism, is the weathervane of these changes.

As human civilization progressed, the harmony between Nature and man captured in literature was irritated. In the fifteenth century, poetry began to decline even in China as a "kingdom of poetry," one piece of evidence to which is that poem writing was reduced to a parading of techniques and learning as is found in many poetic talents, the allegedly pillars of art of poetry.

In contrast, narrative literature such as story-telling scripts, the novel, drama, and ballads to the accompaniment of instruments began to flourish. As a result, the social dimensions or relevance to social men became common themes of representations. The master pieces of this period include *The Romance of West Chamber* by Wang Shifu (?–?), *The Rescue of a Courtesan* by Guan Hanqing (?–1300), *The Golden Lotus* (otherwise known as *The Plum in the Gold Vase*) by Lanling Xiaoxiaosheng (?–?), *The Peony Pavilion* by Tang Xianzu (1550–1616), *Stories to Caution and the World* and *Slapping the Table in Amazement* by Feng Menglong (1574–1645), *Being together after Rebirth* by Chen Duansheng (1751–1796), *The Story of the Stone* allegedly by Cao Xuejin (1715?–1763?) and Gao E (1758–1815?), *Those Scholars* by Wu Jingzi (1701–1754), and *The Legend of a Hero Boy and Hero Girls* by Wen Kang (?–?) What these works present are unexceptionally the lives of different strata of urban dwellers, but "Nature", once a visible presence in the poems of the Tang and Song, shrank dramatically. Though "nature" was depicted, and sometimes beautifully, in the novels and the dramas of the period, it just functioned as the "setting" for the event or a mock-up, thus decreasing in significance so much so that it can no longer be juxtaposed with affairs, human activities, and dispositions.

The replacement of poetry by narrative essays is widely accepted as an epoch-making event in literary history of China and the rest of the world. Undoubtedly, it is not merely about the evolution of genres, but about the fundamental change of the times.

In the Ming Dynasty, as the commodity economy became more active, many changes took place: people's interest began to shift from Nature to the consumption of entertainments such as storytelling and operas. Prolific and popular writers like Feng Menglong emerged. The most popular literary themes were the "rags-to-riches" dream, a life of luxury and nobility, personal bliss like the wedding night or the day of scholarly promotion, etc. Consequently, themes about natural wonders and images such as the awe-inspiring waterfall of Mt. Lushan, the ape shrieking along the Yangtze River, the setting over the waters, and the plume of bonfire smoke rising on the great desert, began to recede in literature and arts.

Some other writers, distaining to flow with the current, wrote in order to satirize and criticize, for instance, the unfair Royal Examination System as done by Wu Jingzi, or the corrupt politics as done by Li Boyuan. As writers, they never returned to Nature in their writings.

An obvious turning point in Tao Yuanming's reception appeared in the Yuan Dynasty.

Partly due to dynasty shifts and the takeover of the throne by ethnic minorities, some scholars and politicians, from a purely political perspective, began to exalt Tao Yuanming's moral integrity as shown by his refusal to serve two different royal courts. Tao was considered merely as a political figure, and his return to the farming life as a last resort rather than a choice out of his free will. For instance, Liu Yin (1249–1293), a poet of the Yuan Dynasty, inscribed in the Painting of "Homeward ho!" that "Tao returns to hoe up weeds, of which Liu Ji'nu is a type." (unpaged)[2] For Liu Yin, what Tao was trying to weed up is not weed itself, but his political rival after whom the weed is named. Wu Cheng (1249–1333), a scholar of the Yuan Dynasty, listed Tao as a gentleman in juxtaposition with Qu Yuan, Zhang Liang (250–186 B.C.), and Zhuge Liang (181–234), who knew well about the sovereign's expectations of them, and who, with rare talents, tried in vain to commit themselves to their states.

Through the Yuan to the Ming and the Qing, there had been ever-increasingly popular politico-moral remarks about Tao. He was depicted as a talented, resourceful, loyal, frustrated, saddened, and lone hero whose ambition was not fulfilled. His return to the farming land was believed to be self-banishment, his farming an act of bidding his opportunities to rise again, and his writing and drinking self-indulgences in order to rid himself sorrows. Tan Sitong (1865–1898), a resounding voice against royal dictatorship and outmoded feudal institutions, believes that Tao and he himself are of the same mind, saying, "A tragic hero, Tao did not mean to be detached from society. It does not make any sense if he is viewed merely as someone enjoying abandon and aloofness" (Tan 375). In fact, Tan, completely identifying himself with Tao Yuanming, abruptly denied that poets such as Wang Wei (701?–761?), Meng Haoran (689–740), Wei Yingwu (737–792), Liu Zongyuan (773–819), Chu Guangxi (706–763), and Su Dongpo (1037–1101) inherited and carried forward Tao's legacy, saying that their writings owe little to Tao.

What is surprising is that after the Ming Dynasty, the focus of the evaluation of Tao shifted from "Nature" to "worldly affairs", though praises for him still abounded. This being said, there had been more criticisms of Tao's poetry.

[2]Liu, Yin. *Anthology of Liu Yin* (IV). Photo copy of the Yuan Dynasty block-printed edition.

In the Ming and Qing, Tao's declining literary status was associated more with the change of economic patterns, values, and aesthetics than with poetry decline. Impacted by the new economic patterns, especially the commodity economy, China's natural economy, which reached its heyday in the Tang and Song dynasties, declined, causing subtle changes in the social structure and cultural mindsets. Proceeding from historical materialism, some historians think that capitalism began to germinate in the Ming Dynasty. Yu Yingshih (1930–) does not agree with this statement, but he believes that the transformation within Chinese society, which started visibly around the time of the Opium War (in 1840), began as in the Ming in an incremental way. For Yu, in terms of philosophy and ethos, this transformation is marked by the "shift of Confucianism in the 15th and 16th centuries" (Yu 189). Yu believes that this shift is expressed in the Confucianists' abandoning learning for commerce, in the people's leaning toward luxury rather than frugality. These trends bear some similarities with scholars' business engagements and people's consumerism today. Yu accounts that in the middle of the Ming Dynasty, some of the large number of businessmen in Shanxi, Anhui, and Jiangsu were at the same time scholars, and some purchased the degrees of "*jiansheng*" and "*gongsheng*", decent ranks certified by the royal court, after they had become rich. These figures become the protagonists in Feng Menglou's popular novels. The distinction between businessmen and scholars was blurred; it was a fashion for scholars to interact with, and integrate into, businessmen. Ironically, scholars, who used to belittle businessmen, now began to sing high praises for them, sometimes for shared interests. Some scholars began to trade their talents for high economic returns. The integration of scholars and businessmen urged "the tradition of valuing agriculture valuation and belittling commerce to adapt to the new social realities" (212). As a result, "the Confucian way gained on new significances because of the involvement of businessmen" (213). Emperor Zhengde of the Ming Dynasty opened up a shop in his royal palace in Beijing. Felt in almost every aspect of the social life then, the Commercialization of scholarship, which explains the popularity of novels and dramas as well as part of the social transformation, accelerated "the turn of Confucianism". (Yu 214) Then and there, the canonized Confucian teaching that "the gentleman values righteousness; the small man treasures profits" almost disintegrated since "Righteousness and profitability were almost one" (219).

Naturally, the pursuit of fortune, comfort, and pleasure had become not only reasonable, but also necessary. *Superfluous Things* by Wen Zhenheng (1585–1645) and *Casual Expressions of Idle Feelings* by Li Yu (1610–1680), covering food, clothing, shelter, and entertainment, are similar to today's consumption manuals. The novels of Feng Menglong indicate that brothels of various kinds and tastes, as part of the mass entertainment industry, could be easily found in big cities. The Confucian admonishment of "sticking to poverty and moral principles" was toppled by the mainstream way of life. The way of life Tao Yuanming represents was no longer encouraged by society. Tao's criticism in his *Lamenting Scholars Fate* that "people are not discrete about integrity and moderation; officials are only driven by the pursuit of higher ranks" was not as welcome in the Ming and the Qing.

As mentioned earlier, Zong Baihua states that people of the Jin discovered Nature out there and their own affections within. Many of Tao's poetic lines like the chrysanthemum-plucking ones are the expression of his affections and his innocence springing from within the depth of his nature. According to Zong, the Jin people were unrestrained, well at ease, and deeply affectionate about their pursuit of Nature, friendship, and philosophy, showing overflowing naturalism both in daily routine and in personality, externally or internally. Unfortunately, commodity worship and the pursuit of luxuries and profits in the Ming and the Qing undermined man's perception of Nature out there and the nature within. In such a social climate, the depreciation of Tao's poetry and prose is understandable, if not forgivable.

Therefore, Tao's philosophy about life and death and vicissitudes, his unwillingness to get rich, and his composure about his destiny became discordant voices in the Ming and the Qing. In fact, Tao's "decline" is already inevitable.

4.4 The Modern Literary Revolution and Tao

The Opium War marks the beginning of China's modern history. After the war, the aging, deteriorating Chinese cultural tradition was severely toppled by western thought. Consequently, the senile China was on the verge of collapse. Confucianism could no longer sustain China, nor could Daoism. The Chinese nation resorted to an embarrassing choice: to learn from the West by absorbing Western culture and following the Western development pattern in order to resist the intrusion of the West and to revitalize itself. During the short span of 79 years from the Opium War to the May Fourth Movement, the Chinese tradition and mindset were changed fundamentally. The threatening national and international realities urged the Chinese scholarship to turn their focus from Confucianism and Daoism to legalism and the thinking of military strategists. As a result, the Westernization Movement, the Hundred Days' Reform was carried out, which reflects the scholarly turn. At that time, in terms of social mindset, "the heroic visa" shone, "the sagely vista" was eclipsed even further, and "the simple vista" became inopportune and therefore unwelcome.

After the 1911 Revolution, especially after the May Fourth Movement in 1919, China, both in terms of the form of government and in thinking, began to break away from traditions, thus commencing its "modernization" and difficult integration into the world. In other words, modern Western thought ushered in the sweeping transformation of China as a senile empire. How would Tao yuanming be perceived and received by the Chinese in such a milieu?

Apparently, Tao's statue saw no big change after China entered into the modern era, at least, not as dramatic as that of Confucius, who, once a revered sage, was dumped and trodden under feet after the May Fourth Movement. Tao was still praised by literary historians and critics as a great poet, but a closer examination may reveal that the nature of Tao's greatness had already changed.

As modernization proceeded, such change became ever-increasingly visible. Metaphorically, the "body" of Tao looked the same, but the soul was already replaced. In fact, Tao's literary spirit, which dwelled in the poetry of the Six dynasties, the Tang and the Song, dissipated; his body was "possessed" by thought and notions that were not his own. This complex situation is similar to what is described as "resurrecting one's spirit by another's body" in *Strange Stories from a Chinese Studio* by Pu Songling (1640–1715). In this sense, Tao was transformed into a different person bearing his name. In contrast, the idea that the soul remains after the body ceases to be is easier to understand.

Hu Shi (1891–1962) and Lu Xun (1881–1936), forerunners of the New Literature Campaign, play a major role in transforming Tao's literary spirit.

Hu Shi advocates "literary revolution" by calling to reform poetry in the first place; therefore, Tao is an inevitable topic. Hu's poetry evaluation criteria are simple: vernacularism and accessibility. For him, it can be a good poem if it is written in vernacular language and readily accessible to the populace; the opposite is also true. His criteria seem to be too rigid. In his *A History of Vernacular Literature*, he follows his criteria through, declaring that whatever is written in vernacular language is worthy of praises and whatever is written in classical Chinese is to be denounced. In Hu Shi, these criteria even apply to those poets who write in both vernacular Chinese and classical Chinese. For him, poems having allusions and those not readily accessible are denounced as strokes of clumsiness.

Hu Shi also thinks that Tao is a great poet, stating that "Among the literature of the Six dynasties, Tao's poetry can be considered as revolutionary because he eradicated the ill inclinations toward Fu (a genre of rhymed prose, tr.), antithesis, parallelism, ornateness, and classicality. He was born a commoner, and returned to his original life after his resignation. … His environment was conducive to the generation of popular literature; his philosophy and learning elevate the *yijing* (the poetic ideo-mood-imagery; the evoked poetic world; poetic conception. tr.) of his poetry. Therefore, his *yijing* is that of philosophers, and his language that of the populace" (Hu Shi 95). Sensitive and observant, Hu Shi discovers the "philosophical significance" of Tao's poetry, noting that "Tao Yuanming is the best example of naturalism. The greatest achievement in his entire life is none other than Nature" (94). He thinks unreservedly that in terms of poetic quality, Tao is superior to Xie Lingyun (385–433), a renowned landscape poet, who, unlike Tao's naturalness and earnest appreciation of nature's beauty, tends to segment natural landscape into symmetrical or antithetical couplets. For Hu, Xie's poems are unnatural representations about Nature, but Tao's poems are natural chantings about Nature; therefore, Tao is a real "Nature poet", to whom many subsequent poets owe much. Hu observes that "Famous nature poets such as Wang Wei, Meng Haoran, Lu You, Fan Chengda, Yang Wanli owes tremendously to Tao's legacy rather than Xie's. … Tao can be viewed as the founder of the landscape school of poetry" (Hu Shi 100). Hu's remarks, sincere, valid and generally accurate, still fail to capture what makes Tao a great poet. The "naturalness" of Tao's poetry observed by Hu Shi is mainly about Tao's language style and absence of ornateness. For Hu, the emergence of Tao represents an unstoppable historical trend, that is, the inclination vernacular literature. Hu does not praise Tao for no reason; he wants to "prove that the vitality of vernacular literature cannot be suppressed for long" (Hu Shi 96).

It is worth noting that Hu's argumentation is not without loopholes since Tao's poetry is not exceptionally simple and as accessible as everyday language, and his prose, in particular, does not completely reject the style of parallelism and antithetical couplets. More importantly, Tao's naturalness is not merely expressed in his diction and techniques, but in his lived "naturalism". Unfortunately, Hu Shi mentions this without any elaboration. If Hu had traced Tao's naturalism to, the metaphysics of the Wei and Jin, to Daoism of Lao Tzu and Zhuangzi, and to the cosmology and ontology conveyed by *I Ching*, what would have been Hu's conclusion? There is no ready answer. However, there is one thing for sure; that is, Hu Shi confesses that he identifies with the literary functionality theory of Bai Juyi (772–846), who advocates that "Inquire about current affairs with conversing with people, and seek truths about state governance when reading history books. One is expected to write to serve practical or functional purposes." Regarding this statement, Hu Shi remarks that "That is to say, literature serves for life and the salvation of individuals and the world; it does not serve for nothing." (313) Underpinning his literary thought, this observation of Hu's captures the purpose of his call for a literary revolution. Obviously, his literary thought is vastly different from Tao's outlooks on life and Nature, including "reading for pleasure", "abandon and aloofness," the resignation to Nature's transformation, the ceasing of the inner world, and eternal composure.

In addition to Hu's observation, Lu Xun's treatment of Tao is worth discussing.

Tao is mentioned more than ten times in *Complete Works of Lu Xun*. In particular, "The Ethos of the Wei and Jin and the Bearing of Medicine and Wine on Writings," "Hermits", and "Untitled (VI, VII)", contain his relative detailed descriptions of Tao. Unlike Hu Shi who focuses on Tao if need be, Lu Xun refers to Tao merely because Tao, in some cases, is relevant to the greater concerns he tries to address. Therefore, Lu Xun's treatment of Tao is not completely free from overdoing or random manipulation. Nevertheless, to some degree, it shows Lu Xun's attitude toward Tao.

In his 800-word discussion of Tao in "The Ethos of the Wei and Jin", Lu Xun, in the better part of it, states that Tao is an impoverished, naturally inclined, peaceful, composed *tianyuan* poet. However, in the latter part of his discussion, Lu Xun changes his tone, claiming:

> Tao is not completely oblivious of, and detached from, worldly affairs … because, there are no tianyuan poets or 'woods poets' whose poetry completely transcends politics. … Tao Yuanming cannot be other-worldly by any means; he is not completely disillusioned by politics; he does not forget about 'death'. … Whatever the perspective, Tao is surely a figure different from what has been perceived and received" (Lu Xun 515–517).

Lu Xun's observations contain nothing new, but his emphasis on Tao's reluctance to bid farewell to politics and social life offers much room of speculation for later scholarship that deals with Tao in Lu Xun's fashion.

In his "Hermits", Lu Xun becomes more critical and harsh about Tao, who, according to Lu Xun, is on par with Lin Yutang (1895–1976) the "hermit", one of Lu Xun's opponents. Lu Xun says that Tao was a renowned great hermit who had servants tilling the land and doing business for him; otherwise, he would have starved to death by the east fence where he plucked chrysanthemums, let alone have drinks inspiring him. Lu Xun's tone is satirical in this case, saying, "Tao accepted the government position for food as he became a hermit for food. If there is no food earned, even hermitage is impossible" (Lu Xun 224). Lu Xun cites a line of Zuo Yan (?–?) of the late Tang Dynasty, reading "failure to seek an official position or hermitage" (224) to show the hypocrisy and treachery of those so-called hermits. These willful satires of Lu Xun, one is certain, are not directed at Tao Yuanming, but at the contemporary "hermits" such as Lin Yutang and Zhou Zuoren (whether Lin and Zhou are hypocritical hermits as Lu Xun describes is irrelevant here). Nevertheless, Lu Xun's skepticism of Tao's hermit life is real.

In his "Untitled (VI)", Xu Lun cites the example of Tao Yuanming to respond to his opponents including Liang Shiqiu (1903–1987), Zhu Guangqian (1897–1986), and Shi Zhecun (1905–2003). Shi Zhecun criticizes Lu Xun's *Ji Wai Ji*, or *Uncollected Works beyond the Collection*, for its lack of discretion in essay selection. Lu Xun replies that previous anthologies tend to include only "Homeward ho!" and "The Peach-blossom Springs," among the more than one hundred pieces of Tao's, and therefore, the readers think that Tao epitomizes unrestraint, ease and aloofness. He continues, "However, in the full anthology of Tao Yuanming, in addition to the lines like the chrysanthemums and southern mount ones, there are also lines like 'Jingwei the bird determines to fill up the sea with small pieces of wood, Xingtian the warrior, after being beheaded, continues fighting without changing his intrepidity. This swaggering and awe-inspiring style proves that Tao's writings are not exceptionally ethereal and aloof" (422). The message that Lu Xun conveys in a roundabout way is that the judgment of a writer should be based on his complete collection of works, rather than on a highly selective anthology, not to mention excerpts. What Lu Xun opposes in this case is one-sidedness in judging people and is basically valid though not without a touch of impulsiveness.

In "Untitled (VII)" Lu Xun refutes Zhu Guangqian's statement that Tao, is great because of his serenity and solemnity all through, arguing:

> No great writer, past and present, is 'serene and solemn all through'. Tao's greatness is attributable to his being not serene and solemn all through. People today believe in Tao's serenity and solemnity because Tao has been disintegrated by anthology and excerpt compilers. (430)

Again, Lu Xun's rebuttal, targeting Shi Zhecun, if not solely Zhu Guangqian, is not invalid in its own right. His problem lies in his conclusion that Tao is great because he is not serene or solemn all through. What Lu Xun reveres is the image of an angry Vajra-like fighter sporadically conveyed in Tao's writings. Nobody has the right to question Lu Xun's personal preference as such, but, when he employs his personal preference as the overriding yardstick to evaluate and rank writers, one may question the legitimacy of this practice. One may ask, considering Tao's sporadic angry Vajra-like style, what has established Tao's status and value in the history of Chinese literature? Is it his engagement in politics and readiness to fight back, or is it naturalism partly shown in his philosophicality, abandon, ease and aloofness?

In fact, the development of poetry in the Tang and Song, especially the poetry that is informed by Tao Yuanming, is an answer to the question. The attempts to alter this course may have to run the risk of indiscretion. One example of this would be Okamura Shigeru (1922), a renowned Japanese Sinologist. In his *A New Reading of Tao Yuanming and Li Bai*, he emphasizes that he follows in Lu Xun's footsteps in rediscovering Tao Yuanming, saying, "The traditional idolization of Tao Yanming is questioned by none other than Lu Xun, the father of modern Chinese literature. Lu Xun intends to offer a fundamental re-evaluation of Tao" (Shigeru 31). Determined to diverge from the tradition and established perceptions of Tao, Shigeru, after a re-examination of Tao, finds that "What is beneath Tao Yuanming is human beings' treachery, willfulness, earthly desires, and deceptiveness" (34). He claims that he has discovered the other side of Tao, who, metaphorically, is an idealized and illusionary Moon. He then declares that "The other side of the Moon is just a boring barren world made up of rocks and sands" (128). In Shigeru's accounts, Tao is depicted as "lazy", "cowardly", "vegetative", "hypocritical", "worldly", "subservient", "shameless", "inner-gloomy", etc.

If Lu Xun had lived to see Shigeru's work, he would not agree with the latter's sensational work and ill-grounded conclusions. In fact, Lu Xun's remarks on Tao, which are different from the widely accepted conceptions though, have something to do with his deliberate "tailoring" to strengthen his verbal fire exchange with his opponents, and to do with personal emotions and impulses. As is known, Lu Xun, a consistent voice for "beating up the drowning dog", "forgiving no one", and "halting so-called fair play," is an intrepid, relentless fighter. Understandably, he treasures Tao's subversiveness and defiance, which is magnified, more than his abandon and aloofness.

In fact, among Tao's contemporaries, there were men of letters who were not serene or solemn but close to fighters. Xie Lingyu, for one, belongs to this category. In Chinese literary history, Tao and Xie are sometimes juxtaposed since the former is the founder of the *tianyuan* school and the latter the forefather of the landscape school. For many years after they ceased to be, Xie's prestige had remained much higher than Tao's. Though Xie and Tao are both attached to Nature, they differ tremendously in dispositions. The former, already a high-ranking official, was eager to climb higher on the political ladder; the latter, a lowly official post holder, was willing and ready to be a commoner. Xie, extremely rich, took delight in displaying his luxuries and extravagances; Tao, impoverished, maintained his integrity and dignity. The former was socially active and this-worldly; the latter was passive and other-worldly. Many of Xie's experiences and anecdotes are vividly accounted in *The Song Annals* by Tuo Tuo (1314–1355) and Ah Lu Tu (?–?) of the Yuan Dynasty and in *The South Annals* by Li Yanshou (?–?) of the Tang Dynasty. According to these historical records, Xie Lingyun, born into an aristocratic family in full plumage, clever, well-learned, and talented, rose to fame when he was still young. Whenever his new poem reached Jiangkang, known as Nanjing today, his readers, the fanatic emperor, and his commoner subjects alike, would transcribe it. Xie can be viewed as innovative in terms of transforming daily routine into aesthetic experiences. According to historical records again,

Xie's chariot was magnificent, and his clothes were always fashion-setters. He even invented "hiking sneakers". Fond of extravagances, He would go on outings with the company of hundreds of his servants, cutting trees to build bridges and roads, creating sensations so big that the intimidated local officials would think that bandits were coming. Talented, defiant, and short-tempered, Xie created more enemies than friends he made in the political circle. Having yeasting political zeal and ambition but no mind given to trickery or deception, Xie would fight back at the sacrifice of his own life when he was challenged. His poetic lines, reading "Zhang Liang avenged himself on the Emperor of Qin for the annexation of his home country. Lu Lian helped to form a coalition of countries against Qin" infuriated the emperor, who issued a decree to have Xie arrested. Before his impending downfall, Xie, unbending as always, allied with local men of valor and strength to fight the royal knights, and, as a result, ended up being killed on the street. Xie's subversiveness and defiance, which Xie explains with his own life, is many times stronger than Tao's occasional anger like that of a Vajra, which is only on paper. However, Xie's extreme subversiveness and defiance undermines his naturalness as a landscape poet; therefore, his poetry is sometimes plagued with ornateness, extravagance, and, a trace of indoctrination. For this reason alone, his poetry is inferior to that of Tao in terms of naturalness and integrity. In addition, his poetry is not as impulsive or unique as that of Ji Kang (224–263) and Bao Zhao (415–466).

Partly due to the unparalleled status of Hu Shi and Lu Xun in Chinese literature studies, and partly due to the changing zeitgeist of modern times, the tone for the evaluation of Tao Yuanming in the post-May Fourth Movement period was set, highlighting "vernacularism", "grassroots", "subversiveness", "defiance", and "the angry Vjara style". After the power shift in 1949 in the mainland of China, the evaluation of Tao by Lu Xun and Hu Shi (usually labeled as a political dissident) is passed on in university textbooks from one generation of students to another, thus establishing itself as the predominant, if not the only, criterion to judge Tao.

Undoubtedly, Hu Shi's and Lu Xun's evaluation of Tao exactly represent their ideals of "modern literary revolution." Informed by the Confucian scholarship's shift to practicality in the Ming and Qing, and informing the guideline that literature should serve politics, which is a mentality of class struggle, their judgment, buttressed by Hu's instrumentalism and pragmatism as well as Lu Xun's invincible spirit to fight, urges the literary circle and scholarship to construct Tao's spirit in modern times.

Li Changzhi responds to this trend with his 1952 book titled *A Biographical Study of Tao Yuanming*. In the foreword, he confesses that he attempts to reinterpret Tao Yuanming from the perspective of political stance and ideology, and to realize Lu Xun's unfulfilled wish of "discovering a real Tao that is different from previous observations" (Li CZ 1).

In his elucidations, Li, instead of drawing on his real bodily and spiritual experiences gained in his childhood and early manhood from reading Tao's writings, resorts to the theory of class struggle and ideology to analyze Tao and his writings. This is sad enough. Worse still, Li's mind is set on diverging from previous observations, and, worst of all, he succeeds in "constructing" a brand-new Tao Yuanming. According to Li's narrative, Tao, born into a declining official family and once a landlord, was engaged in farming and close to real farming people, thus becoming a spokesman for grassroots people. "Among all Chinese poets, there is no one like Tao who feels for labor and is physically engaged in it. It is in this sense that Tao is great" (Li CZ 129). In his work, Li describes Tao as a Confucian scholar who never forgets political struggle. In Li's eyes, Tao returned to farming life just on the spur of the moment and out of helplessness rather than out of free will; his choice is "not as simple as he himself describes in his poetic line 'gladly accept Heaven's will' (69)." Li also claims that Tao's silence is to oppose politics, and his aloofness to oppose realities. Determined to implement Lu Xun's impulsive comment that Tao is very concerned with social realities and has a strong spirit to fight, Li finds that Tao's alleged great grandfather Tao Kan (259–334) carried blood of the atrocious and belligerent *xi* tribe. Li assumes that Tao's line "The harvest is not going to the royal court as a tribute," which appears in *The Peach-blossom Springs* is identical with the old uprising peasants' slogan "Follow Chuangwang the rebel leader; pay no tributes" (Li CZ 138). This shows that Li goes even farther than Lu Xun in "deforming" Tao Yuanming because Li is particularly subject to the political climate of his day. Ironically, Li's work, far from being a success in China, became a sensation in Japan and has influenced, among others, Shigeru, who, in turn, tried to discover the "dark sides" of Tao, as mentioned earlier.

This purely political, class-struggle-style critical paradigm culminated during the 10-year-long Cultural Revolution. The merciless criticism of the zealous public aside, many well-trained and well-learned sympathetic scholars, subject to the current of the times, and to the political tension, had to diverge from their true intensions and exhaust their derogatory, insulting words for Tao. Consequently, Tao's return to *tianyuan*, his contentment with his poverty, his drinking, and his writings such as "Excursion to Xiechuan" and "The Peach-blossom Springs" were exceptionally negated. During this period, the status of Tao, now a "right-wing" poet, could no longer be maintained in the history of Chinese literature. If it could be maintained at all, it is because that Tao could function well as a negative example.

4.5 The Death of Tao

There is no consensus on when Tao was born, but it is recorded in *The Jin Annals, The South Annals,* and *The Song Annals* that he died in November of 427. As mentioned earlier, most poets of all generations after Tao are informed and nourished some way by him. The plethora of traits such as naturalness, sincerity, aloofness, philosophicality, dignity, contentment with poverty, independence, resignation to Nature's transformation, and his naturalism demonstrated by Tao's writings and his personality, has been nourishing Chinese literature and culture. In this sense, Tao does not cease to be; his spirit is still alive, and his writings, which symbolize his life, are still alive.

What I mean by "the death of Tao" here refers to Tao's second death. Tao is dying, because his spirit, philosophy, and writings, which symbolize and convey his spirit and philosophy, are fading. The impending death of Tao in today's world is real, devastating, and symbolic.

Since the 1950s, though Tao has remained a topic at academic conferences, its influence has been declining in people's everyday life and the literary circle. If he appears at all, his image is either unwelcome or deformed.

In July, 1959, Mao Zedong (1893–1976) composed a poem titled "Climbing Mt. Lushan," the most interesting two lines of which read "Where has been Tao the head of Prefecture? Is he plowing in the Peach-blossom Springs?" Some authoritative critics at that time interpret the lines this way: "Tao has been out of date. Could he be farming in the Peach-blossom Springs? It was impossible and that was just his reverie. The countryside today is different from the Peach-blossom Springs, and intellectuals today are also different from Tao as an ancient intellectual" (Zang and Zhou 37).[3] This statement implies that the chapter of Tao in Chinese cultural history has been turned over. In his *Notes on Reading Works of Mao Zedong*, Chen Jin (1958–) releases that Mao Zedong once confessed to Lu Di (1931–2015), his classical literature reader-in-waiting, that "Even real hermits are not to be praised. Tao's becoming a hermit, for instance, is excessively exalted" (Chen J 86).[4] According to Chen Jin's interpretation, Mao's mentality may be like this:

> Is it not the responsibility of the scholar in every society to become a better man, raise a decent family, govern the state, and make the world a better place? It is fine if you wish to become a hermit. But Tao was serious about farming as a hermit, which is a waste of educational resources. Worse still, Tao forgets his greater social responsibilities and historical missions. If all learned intellectuals retreated to the woods or mountains and people abandon their wisdom as Lao Tzu and Zhuangzi advocate, how would human civilization progress? (86)

Chen's interpretation of Mao's mentality seems reliable. In fact, Mao's remarks about Tao in his poem mentioned earlier set the tone for the evaluation of Tao at that time: a passive, conservative, rightist intellectual who runs counter to the current of the times.

In the 1960s, a literary–political incident regarding Tao Yuanming happened, bearing witness to Tao's harder destiny and the cruelty of class struggle. The incident was caused by "Tao Yuanming Writing an Elegy," a novelette by Chen Xianghe (1901–1969), a writer of sincerity and introversion, an orchid lover and admirer of Tao Yuanming. "Tao Yuanming Writing an Elegy" was published on Issue 11 of *People's Literature* in 1961 and reviewed very positively.

By today's standards, the novelette is interesting and smooth, reminding people of the charm of the literature in the 1930s. More importantly, it captures Tao's

---

[3]Zang and Zhou (1990).

[4]Chen (2009).

spirit with a touch of sentimentalism in philosophicality, and a touch of self-relief in cynicism. The only flaw, if any, is that the narrative of Tao's experience is too subject to historical accounts. However, this guarantees that the Tao presented in the novelette is very close to the Tao in the real world, thus having a strong sense of reality.

The gist of the novelette is as follows:

During the autumn of 428, after a not very pleasant interview with monk Hui Yuan of the Donglin Temple in Mt. Lushan, Tao Yuanming walked about 13 miles back to his home and had a sleepless night. On the next day, soothed by familial bliss, he had a sense of relief. He told his family about the secularity and affectation of celebrated Buddhist Monk Hui Yuan, the arrogance of the noble Wang Hong and Tan Daoji, the shallowness and mediocrity of Liu Yimin and Zhou Xuzhi as famous scholars; the kind-heartedness and indecisiveness of his friend Yan Yanzhi, his contempt of the newly crowned Emperor, and his identity with the virtue of Ruan Ji, a sage of the previous dynasty. Inclusive as it is, the novelette is natural, pleasant, and smooth, thus presenting Tao's spirit and view of life and death without any touch of moralizing. Then and there, Tao was moved by the elegy he was composing for himself, "lamenting his life, present and past" (Chen XH 163). In addition to Tao's spirit and legacy, especially his virtue, integrity, and aloofness, what the novelette also expresses are the depression of Chinese intellectuals caused by prosecutions and persecutions of various kinds during the Cultural Revolution, and their wish to seek self-relief through creative writings. The novelette evokes a real Tao Yuanming as scholars perceive, and the intellectuals' identity with Tao in the depth of their souls. It is very rare in the post-1949 Chinese literature and should be viewed as proof to the remaining though dying spirit of Tao.

However, the hope for Tao's revival was nipped as the criticism of "Tao Yuanming Writing an Elegy" began in 1964, when some political authorities decided that the novelette was "unhealthy", "gloomy and dark," "inclined toward a gray view of life," and "full of the lamentations and reveries of the declining capitalist class." This was not only the judgment of the novelette itself, but also of the political aspect of Tao Yuanming. Shortly after the criticism, Yao Wenyuan (1931–2005), one of the "Gang of Four", an absolute authoritative critic, states in an article that some CPC members are not enthusiastic about revolution, but very enthusiastic about the tastes and pleasures of Tao Yuanming's life. During the Cultural Revolution, the setting of Chen Xianghe's novelette was juxtaposed with "the Mt. Lushan Conference" (also the setting for Mao's Poem "Climbing Mt. Lushan") and was declared to be a voice for the grievances of the downed counter-revolutionary capitalist class, to be a sort of instigation to call for "fight back", and to be a poisoned arrow at "the CPC and the proletariat dictatorship." Consequently, Chen Xianghe died on the way to yet another round of "public condemnation" on April 22, 1969. Ironically, "Tao Yuanming Writing an Elegy" became an elegy for himself.

After the Cultural Revolution, as scholarship resumes, studies of Tao Yuanming become active and fruitful, as shown by the emerging works on Tao, including Yuan Xingpei's *A Study of Tao Yuanming*, Gong Bin's *A Biographical Study of*

Tao Yuanming, Hu Bugui's *Notes on Reading Tao Yuanming*, to name just a few. However, Tao's reception was faced with another grave challenge at the turn of the twentieth century, though not necessarily in Chinese scholarship. China is shifting from a traditional agricultural society to an industrial one and from a planned economy to a market economy. Economic development has become an overriding criterion to judge all undertakings; the attention of the public has been drawn to making fortune; consumerism and the law of market economy have dominated public and private domains, including health care, education, press, religion, and spirituality, which are traditionally considered as cultural and wellbeing undertakings. Quantity, royalties, box income, etc. have become the most telling yardsticks to evaluate works of literature and arts. Scholars are vulnerable to this trend. In *The Philosophy of Money*, what Georg Simmel (1858–1918) says about money, such as "non-personality", "colorlessness", "indifferent quality," and "flatness" are also features of society today, where money has become an almighty, ubiquitous presence, and where human life has been depersonalized and rendered colorless.

Against this background, the countryside is being urbanized, agriculture is becoming the produce processing and manufacturing industry, peasants are becoming migrant workers in the city, and the *tianyuan* landscape is disappearing. Worse still, what underpins the nation's education from kindergarten to university is a pragmatic notion of education. Educational governance has been bureaucratized and marketized. Becoming businesspeople or officials have become the ideals of many young students. It has become a social norm that many masters and doctorate holders dream about becoming "public servants", or government officials, and scholars just bid their opportunities to rise, financially or in social ranks, even at the sacrifice losing much of their independence, integrity, and dignity. As peasant workers swarm into the city for better job opportunities and beyond, how can one still chant "Thou return! Thy field has weeds growing!"? As thousands of young promising scholars compete for a single available lowly government position, how can one still stick to Tao's principle of not bowing for government salaries?

Even Tao's so-called "grassroots nature" and "subversiveness" as observed by Hu Shi and Lu Xun are in plight now. After the Cultural Revolution, the nation's decision makers decided not to continue the guideline of class struggle and make "the spirit to fight" an inseparable part of every individual's political life, though "classes" never cease to be, nor do "struggles" stop. The slogan "class struggle in mind for ever" has been replaced by "Stability is everything." Tao's so-called subversiveness, which was employed to serve political purposes, has lost its vitality and legitimacy now. "Grassroots nature" and "popularization" espoused by Hu Shi seem to be more complex, because the popularization of literature and arts, and the transformation of aesthetic experiences into daily routines are not to be realized by writers, aestheticians, artists, or even political leaders, but by capital and technology represented by so many modern gadgets and conveniences such as the mobile phone and the internet. It have been an established fact that those who are most pleased by popularization are not the populace but business owners. If "The customer is God," what is there driving thousands of millions of gods? In a sense, popularization is identical with the transformation, if not manipulation, of the populace. In today's booming market economy, cultural undertakings are becoming industries or enterprises, those powerful capitalists have more than one way to corrupt or silence the unlucky grudging cultural elites. Unexpectedly, even Tao Yuanming himself has also failed to escape the net of popularization cast by powerful entrepreneurs, "cultural elites," and pop stars in the cultural industry.

An example would be *Secret Love in Peach Blossom Land*, a sensational drama directed by Taiwan-based director Wei Shengchuan. It was a huge success on the mainland of China, both in terms of box income and social reception. With costume drama *Peach Blossom Land* interwoven with fashion drama *Secret Love*, and a perfect blending of the ancient and the modern, sadness and rejoice, the drama makes the audience cry now and laugh then. The director's and script writer's talent help to deconstruct Tao Yuanming and the Peach-blossom Springs. What are registered in the audience's minds are the images of "Old Tao", "Peach Blossom (Old Tao's wife)" and "Boss Yuan" instead of Tao Yuanming and the Peach-blossom Springs as we know.

An even less serious example would be Teng Ko Erh, an inner-Mongolian composer–singer, whose "brainwashing" song "the Peach-blossom Springs" was also a sensation. In his music video, the Peach-blossom Springs becomes a night club where there are numerous beautiful girls, and the fisherman becomes a playboy. Such abnormal and insulting performance was staged at the Beijing Spring Festival Gala in 2014.

An internet search for the "Peach-blossom Springs" would activate hundreds of property advertisements because real estate developers like the idea of naming their properties "the Peach-blossom Springs." The real Peach-blossom Springs as envisaged by Tao Yuanming has been replaced by the images of the real estate bearing its name. Obviously, such tangible peach blossom colonies are even more alluring to people in the real world.

Therefore, not only Tao, as a man of bone and flesh, who died about 1,600 years ago, but also his spirit, are being abandoned today. It seems that Tao, now simply an outsider in twenty-first century China, has lost his s legitimacy, marking the plight, if not death, of his literary legacy and naturalism. It is in this sense that Tao is an "Other" in contemporary China. The death of Tao is worth contemplating. According to Derrida's theory of specters, only after his second death, can Tao become a real "specter", heralding a possible resurrection.

Chapter 5
The Specter of Tao and the Plights of Contemporary Human Existence

The instrumental rationalism of the Enlightenment disintegrates the paradigm of the universe characterized by "knowing the bright but cleaving to the dark" and therefore undermines the organic connection and dynamic balance between the different aspects of the earth's ecosystem and impacts the unity of matter and spirit, civilization and Nature, knowledge and faith, sense and sensibility, men and women, the country and the city, technology and culture, progress and conservatism, power and vulnerability, benefits and conscience, clarity and vagueness, quantity and quality, etc. It seems that *Lao Tzu* offers a prophetic description about the consequence of such disintegration:

> Were it not so limpid, the sky would soon get torn.
> Were it not for steadiness, the earth would soon tip over.
> Were it not for their holiness, the spirit would soon wither away.
> Were it not for this replenishment, the abyss would soon go dry.
> Were it not that ten thousand creatures can bear their kind.
> They would soon become extinct.
> Were the barons and princes no longer directors of their people,
> And for that reason honored and exalted, they would soon be overthrown. (85)

Is this not a vivid description of the impending eco-crisis?

As discussed earlier, it is the Chinese tradition to identify Nature as "Heaven" and to believe that the origin of all things is primordial "*qi*" (air; life-force). However, both "Heaven" and "*qi*" are transformed by instrumental rationalism or, specifically, by man-made gases like Freon. Greenhouse gases intervene with the original *qi* and "Heaven" above, partly contributing to global warming, to the extinction of many species, and to the thinning of the ozone layer over the Antarctica. "The air" as once breathed by Tao in his poetry is no longer fresh; Tao's "Heaven" is deprived of its aura of holiness; his winds and clouds, now containing suffocating construction and even nuclear dust, are no longer completely natural presences; his birds (especially swallows) are rarely seen, if at all; his field is rendered silent by fertilizers and pesticides; his line "The caged birds are attached to their original woods" loses its soothing warmth since magpies in many

S. Lu, *The Ecological Era and Classical Chinese Naturalism*,
China Academic Library, DOI 10.1007/978-981-10-1784-1_5

"economic development zones," having difficulty finding twigs, use thin iron wires and debris to build nests to protect their nestlings and themselves.

Natural catastrophes aside, the consequences of the mistakes humans have made over the past three hundred years and are still making now, in politics, economy, culture, and education, never stop. Though barefaced racial discriminations are curbed now, the "jungle law" and the "bulling nature" of international interactions have not changed fundamentally, if not for the worse. The hegemony, domination, and willfulness of some powers urge those backward small and weak countries to "rise" in terms of power. From the UN to grassroots organizations, in many cases, truth does not speak louder than power, and therefore, the tender and the weak have no secure footing. Besides, humans are faced with the risk of nuclear proliferation and even nuclear wars, which would eventually destroy the ecology and entire human civilization.

Such being the world without, things are no better in the world within.

At a remembrance meeting in honor of his fellow citizen musician in 1955, Heidegger lamented that nature has become a huge gas station, modern technologies, and power, that it will not be long before life can be controlled by chemists since organisms can be disintegrated and assembled at will, that everything falls into planning and calculation, that the mysterious world has become a technological world, that humans will negate and abandon what essentially makes humans thinking organisms, and that human life will be uprooted from the earth and thrown into a totality of thoughtlessness and rootlessness (Sun 1234–1241). In fact, natural crises and fiascoes are also spiritual ones. Therefore, in today's world of high technology, affluence, and luxury, there has come an unprecedentedly barren, arrogant, and disorderly era in terms of public good and private morality.

Anthony Giddens, affectionate about the world as he is, is confused, asking, "Why, in any case, do we currently live in such a runaway world, so different from that which the Enlightenment thinkers anticipated?" ("CM" 151). He is not sure about whether it is caused by "design faults" or "operator failure." He wants to know "How far can we-where 'we' means humanity as a whole-harness the juggernaut, or at least direct it in such a way as to minimize the dangers and maximize the opportunities which modernity offers to us?" ("CM" 151).

Regarding Giddens's question, many thinkers before and after him have tried in vain to offer a mature and effective solution. Here in this case, I want to trust the mission of salvation to Tao Yuanming, or specifically, the specter of Tao Yuanming, the avatar of Nature and freedom.

According to Derrida's specter theory, the dead are even more powerful than the living, and "The specter … is also the impatient and nostalgic waiting for a redemption" (Derrida 170). It is expected that the tender and feeble specter of Tao Yuanming would break the tight "cage" of modern society and bring Nature and original simple happiness back to human beings. Admittedly, this may be purely an expectation in the nature of literary whimsicality, but, barring this, on what can humans rest their hopes of salvation?

5.1 The "Specter" and Derrida's "Specter" Theory

The specter that appears in Chinese folk tales and literary works is similar to the English language specter or German Gespenset. For instance, the specter of General Guan Yunchang in *The Romance of the Three Kingdoms* by Luo Guanzhong (1280–1360), who yelled "Return my head" and asked his sworn brother Liu Bei to avenge, is similar to the specter of the King of Denmark in Shakespeare's *Hamlet*, who asked his son to take revenge for him. The specter is the apparition of the spirit after the body ceases to be, and it is a misty, dreamy, thingless presence. It is believed that it dwells in the depth of human heart and spirit when the body lives, guarding the body's vitality and existence. After the body dies, the specter comes from within the body and becomes more powerful than the body itself, flexible, changeable, eternal, and almost invincible. Presumably, the image of the specter as such inspired Derrida in his establishing specter theory in later years.

Regarding the meaning of "specter," Chen Xiaoming (1959–) says the following:

> Among the words "specter", "spirit" "ghost", "phantom", "demon", "monster", "nymph", "fairy", "vampire", "poltergeist", etc., "spirit" is abstract, but it still has an image. "Specter" is the most abstract, and it reveals itself only sporadically, and only by virtue of other carnal entities. Therefore it is close to hunpo (魂魄) as the Chinese perceive it. ("BLD" 507)[1]

The specter in the philosophical sense as defined by Derrida is paradoxical as it is in literary works. Derrida believes that the specter itself is a paradoxical presence, because it is neither spiritual nor carnal, neither unreal nor real, and neither visible nor invisible. For Derrida, it only appears after the body dies, but unlike its former self when dwelling in the body, it is now changeable, accumulative, and reproducible. Derrida states that "the phenomenal form of the world is itself spectral" and that "the phenomenological ego … is a specter" (Derrida 169). Interestingly, even "humanity is but a collection or series of ghosts" (172). The specter or ghost in the Derridean sense is more like spirit or spiritual legacy though he deliberately distinguishes the specter theory from all previous theories about "spirit." Derrida says, "The specter is not only the carnal apparition of the spirit, its phenomenal body, its fallen and guilty body, it is also the impatient and nostalgic waiting for a redemption" (170). If "specter" is considered to be similar to, or identical with, "tradition" or "spiritual legacy," it is not necessarily the dead past or simply passive; in fact, it is alive and has a mission. Derrida's specter theory proceeds from but transcends "spirit" or "spiritual legacy." Derrida argues that "If we have been insisting so much since the beginning on the logic of the ghost, it is because it points toward a thinking of the event that necessarily exceeds a binary or dialectical logic" (78). In Derrida, "specter," an everyday word, is infused with

[1]Chen (2009).

philosophical intension, and therefore, his specter theory is of methodological value.

In terms of nature, the specter in the Derridean sense has many similarities with the Dao of Lao Tzu and Zhuangzi. Zhuangzi says:

> Emptiness leaves no trace; change signifies no form. There is no life or death. [D]ao coexists with the heaven and the earth and communicates with both the internal and the external. It goes nowhere in a trance and comes from nowhere in a trance. It contains everything and yet does not know its final destination. (601)

Unlike in Western philosophy where the nature of the subject tends to be revealed and highlighted, traditional Chinese philosophy usually resorts to vague language to blur the distinctions between things seemingly different, and in so doing, people may be able to see more of what is otherwise invisible, for instance, the specter. In his mock dialogue with Derrida, Shang Jie says that only by breaking away from what is habitual can one see what is invisible such as "bodies without bodies and things without thingness" (Shang "MD" 262). This is perhaps where "primitive thinking" or "mythological thinking" may be better than scientific, logical thinking.

The specter theory and witchcraft or necromancy as methodology remains a long-established tradition among many peoples, though it has almost been "disenchanted" and "exorcised" first in the West and then in the East by science and technology. Derrida as a post-modern philosopher resorts to the specter in order to explain the many puzzles in the spiritual realm of humanity. For the deconstructionist Derrida, the more primitive a language is, the more natural and truer it will be. In fact, the roots of philosophy are usually in primitive religion, mythology, or wizardry.

The specter of Marx fascinates Derrida profoundly. Marx (1818–1883) transforms Hegel's philosophy of rationalism into a philosophy of practice. In Marx's philosophy, Hegel's "absolute idea" is replaced by productivity and production relations, Hegel's phenomenology of mind by the law of class struggle, and his kingdom of freedom by the ideal of communism. It seems that Marx, who turns Hegel upside down, still fails to escape the Western thought pattern, especially the specter of "logos." More precisely, the specter of logos provides logical support for Marxism.

No philosophy has brought about as fundamental a change worldwide as Marxism has ever done. However, the philosophy of communism is now in plight. Francis Fukuyama (1955–) proposes his "the end of history" assertion, claiming that Marxism is phased out, and human history would end in the form of American-style global capitalism.

Derrida's *The Specter of Marx* is, in fact, a response to Fukuyama's "the end of history" assertion. For Derrida, what is phased out is the giant Soviet Union and the community of East European socialist countries rather than Marxism or Marxist spirit. Derrida is convinced that the decease of the body and any attempt to eradicate Marxism would end up evoking Marxism as a specter and that the specter of Marx is even more powerful, subtle, and resilient than Marx himself.

Interestingly, it is a specter as in "A specter is haunting Europe—the specter of communism," the opening sentence of "The Communist Manifesto" that led to a series of sweeping communist movements.

Not subject to the constraint of real bodies, the specter is eternal and omnipresent and can be evoked again and again. Metaphorically, it is "a case of unweaving the 'old cloth' yet again" (Sim 47). Therefore, "Marx is too deeply ingrained in our cultural heritage to be dismissed" (Sim 45). Even today, people all over the world are still connected to Marx and Marxism in one way or another.

5.2 Tao as a Specter in the Kingdom of Poetry

It is safe to say that China is a kingdom of poetry as Germany is one of the philosophies. Though every nation has its own great poets and poems, perhaps no nation in the world owns so many excellent poems and poetic pieces as China, where the spirit of poetry is still vital. China has well over 100,000 poems available from the eleventh century B.C. to the late Qing Dynasty, exclusive of ballads, folk songs, etc. Among its galaxy of poets, who can China present to the world for the sake of the salvation of Nature and the ever-increasingly barren inner world? Undoubtedly, that poet should be nobody other than Tao Yuanming.

As mentioned earlier, Tao died about 1600 years ago, and his body, as Tao himself describes, is formless and one with the mountain. In fact, Tao is not very sure about whether human beings have souls or not. In his "In Memory of My Sister Mrs. Cheng," Tao writes, "Who is to mourn your wondering soul? … If the deceased have consciousness, we shall meet some day in the graveyard" (251). Tao also tries to imagine what happens after death, saying the following:

> In the past I sleep in spacious halls,
> But now I lie in fields where wild grass sprawls.
> Once I leave my home in dreamless sleep,
> My soul alone returns when night is deep. (211)

It seems that Tao does not deny the existence of souls. However, most psychologists today would refuse to acknowledge eternal souls, though Jung is a rare exception, who believes that the soul can live independent of the body.

The question arising here is whether Tao can be a specter in the Derridean sense of the word, or, more precisely, what kind of specter will Tao become? To answer this question, it is necessary to revisit Derrida and compare Tao against the former's standards.

Firstly, in Derrida's account, a specter does not live in the same way as the body does, "nor does one see in flesh and blood this Thing that is not a thing" (Derrida 6). The specter lives after the body dies, and for this reason, it is eternally there. In the history of Chinese literature, Tao Yuanming is often referred to as "a thousand-year-old man" because his literary legacy remains. In the poems of

Su Shi, Chen Yuan (1607–1145), Ge Shengzhong (1072–1144), Huang Tingjian (1045–1105), Qiu Wanqing (?–1219), Zhu Xi (1130–1200), Wang Ruoxu (1174–1243), Yuan Haowen (1190–1257), etc., Tao is described as "eternal," "present for one millennium," "a poetic soul of one thousand years," or "a soul mate one thousand years ago," all pointing to the eternity of Tao's poetic spirit. Tao becomes much more influential after death. The case can be described by "A specter is haunting Chinese literature—the specter of Tao Yuanming."

Secondly, another character of the specter is that after its physical carrier is gone, it becomes some sort of spiritual phenomenon, and "The Thing [Chose] haunts, for example, it causes, it inhabits without residing, without ever confining itself to the numerous versions of this passage" (Derrida 21). Mistlike and residing nowhere, Tao Yuanming is also such a presence, going beyond time and space. He is never confined to the imitations or interpretations of his followers. Regarding Tao's residing, some remarks are worth noting. The remarks by Ye Mengde (1077–1148), Liu Yin (1249–1293), Yu Ji (1272–1348), etc., all mention the spiritual presence of Tao. In particular, a poem by Yang Wanli (1127–1206) contains a good description of the state of Tao's being: Tao forgets his words after having got the meaning, but his meaning lies beneath. "When asked where his true meaning resides, I would say it is like an image in the mirror or the sky in the water." ("Inscription on the True Meaning Pavilion of Liu Defu") Coincidentally, this narrative about Tao just agrees with Derrida's definition of the specter: something between being and non-being, and the real and the unreal. "When Qi, or the vital-force, is transformed into things, it is something out of nothing, a wondering spirit is nothing out of something. … The bone and flesh only return to the dust … but the wondering spirit is omnipresent" (Gu 35–36). Gu Yanwu's observation is a good footnote to the specter.

Thirdly, Derrida believes that the specter is a form of inheritance, and "Inheritance is never a given; it is always a task" (Derrida 76). It is "the impatient and nostalgic waiting for a redemption" since "a ghost never dies, it remains always to come and to come-back" (170–123). This is to say, a specter does not always belong to the past; it comes back any time, informing future. In this sense, it can be a psychological tension between the past and the present.

After a visit to Tao's former residence, Bai Juyi writes: "I was born five hundred years later after you were born. Whenever I read your 'Biography of Five-Willow Gentleman', you appear in my sight and thought." Xin Qiji (1140–1207) writes in the same vein: "Only in my old age did I get to know Tao. I see him often in my dream. I believe Tao is not dead, since his moral is still there. Tao knows me, past and present." The experience of Chen Yuan (1607–1145) is even more interesting:

> Yuanming has returned to the earth.
> His poetry remains, novel and tasty.
> Whenever I roam in the wild,
> And, eyebrow raised, I will see him.
> Who says the ancient man is far away?

He is not departing or staying.
Read his poetry and digest his words,
And you will know where he is. (Chen Y Unpaged)[2]

Does it not seem that Tao, like a specter, or as a specter, is coming to us in the field?

Finally, Derrida thinks that the specter is accumulative and reproducible. It means that a specter can give birth to another "self" through "the incarnation of the spirit," and "it is in turn negated, integrated, and incorporated by the very subject of the operation" (Derrida 158). Analogically, the reproduction of a specter or the specter itself is like a reincarnated soul boy in the tradition of Tibetan Buddhism. Or, in folklore, it is the returning of the soul by a possessed body. An example of this would be Su Dongpo, who believes that he is reincarnated Tao Yuanming. He confides to others, saying "I realize in my dream just like I become sober from my drunkenness that I was Yuanming in my previous life" (Su Shi's poem). It is recorded that on a day in 1095, Su Shi, drunk and with a full stomach, sat still in complete forgetfulness and in mistiness saw Tao Yuanming. After he woke up, he transcribed Tao's "A Man in the East" and gave it to his son Su Guo, saying "I was Tao; Tao is me." As is mentioned earlier, Tao has nourished many poets after him, and in this sense, those who try to imitate him or inherit his legacy are possessed by Tao Yuanming, or become reincarnated as Tao Yuanming. Or ontologically again, if Tao were Buddha, and those who are exposed to his light and spirit would become bodhisattvas.

Tao's "The form, the shadow, and the Spirit" itself can partly explain that Tao now is a specter. Simply put, "the form" is the real body of bone and flesh living in the real world. "The spirit" refers to his resignation to Nature's transformation, and it is a spiritual form of Tao's nature and poetic conception. "The shadow," as an intermediator between the body and the spirit, is both the mirror image of the body and the carrier of the spirit. For Tao, his "shadow" is exactly his poetry and prose, in which the readers will "see" Tao coming to them. Our discussion should not be hindered by the debates over materialism and idealism, or by the notion that "if the body dies, the spirit vanishes," which is established in the Jin and the Six dynasties. After many centuries' reception and "baptism," Tao still remains vital and is reincarnated in many men of letters. He already transcends the constraints of the form and the shadow and therefore enters into a higher state of being, or, as Zhuangzi puts it, "a spirit beyond the spirit." Thus, he becomes a specter in and of Chinese literature.

Our discussion here is not intended to be a study of Tao's "form," which belongs to the category of biographical studies, or to be a study of Tao's shadow, which falls under the category of textual studies. Instead, it is aimed at exploring the Tao's specter in the Derridean sense of the word.

---

[2]Chen, Yuan. *The Collection of Motang* (V.) Photocopy of the rare ancient edition.

Tao has become a specter, but that does not mean that everybody can "see" it. "Only those minds that are subtly trained may be able to see the invisible in the process of reading" (Shang "MD" 259). Similarly, Chen Xiaoming argues that in the context of the literature, the apparition of a specter is also a form of the apparition of tradition. For him, tradition can only be spectral and reveal itself like the apparition of a specter. It is not to be found in textual details, because, in most cases, it is just a spirit, a spectral spirit ("FE" 6–28). Now, the real question is how can one "see" the specter of Tao and evoke it in this materialized earthly world?

Now that Tao is a specter, his perception and reception is not "a given," or "intended"; it is unintended. In other words, what one may receive from Tao is some sort of spectral wonder. In this sense, Tao's influence on Chinese literature and culture dwells predominantly in literary works, especially those imitations.

On the inheritance of Tao's poetic spirit, Shen Deqian (1673–1769) says the following:

> Tao's poetry finds its way into Meng Haoran's poetry, but the former's poetry always has something out of reach. Among Tao's followers in the Tang Dynasty, Wang Wei has Tao's naturalness and purity, Meng Haoran has Tao's ease and aloofness, Chu Quangxu has Tao's simplicity and plainness, Wei Yingwu has Tao's calmness and serenity, and Liu Zongyuan has Tao's integrity. All of them are similar to Tao in nature, each in a different way. (164)

My understanding is that Tao's poetry conveys or constructs a spiritual world that is not readily accessible to others and that is exactly the spirit of Tao's poetry. Poets are not expected to be the exact imitators of Tao since Tao's "spirit" is not something inscribed on stone. Once it finds its way into the realm of collective unconsciousness, Tao's poetry is no longer itself, but rather a deviation of itself. The evolution of Tao is a process of "difference" in Derrida's term, "difference" being used to describe that meaning, in its generation and flow, is open to all possibilities, and moves endlessly in ambiguities, thus becoming the soil for more meanings. All the imitations of Tao's poetry are only meanings that Tao's poetry generates in a way of "difference." They, like segments though, are reminiscent of Tao's spirit. Regarding the "difference" of Tao's spirit in Chinese literary history, two "subtleties" should not be overlooked. The first subtlety is about the differentiation of Nature and institutions, and the other subtlety is about the "vista of life" in Jin Yuelin's term as mentioned earlier (Jin identifies the heroic vista, the sagely vista, and the simple vista). Ultimately, the core of the both points regards man's attitude toward Nature, or, more generally, the relationship between Nature and man. Tao's status in Chinese literature history can be partly attributed to his inclination toward Nature and defiance of institutions, and to his simple vista of life instead of a heroic vista or a sagely one. Besides, he expresses this with unrivaled talent in his poetry and prose.

Due to their different social status, life experiences, and personal perceptions of Tao, those who try to follow Tao's life style have to choose between engagement in society or detachment from it. In fact, there are far more people who decide to commit themselves to society than those who are determined to retire to a solitary life. Tao has the most followers in the literature, which, in turn, is hostile to reality.

Most of the Tao's followers begin to feel for Tao and understand Tao's inner world only after they have experienced numerous vicissitudes in their life. This is the case with Wang Wei, Chu Quangxu, Wei Yingwu, Liu Zongyuan, Su Dongpo, and Xin Qiji. Well versed in classics in their childhood, and talented, they decided to devote themselves to their countries. Holders of relatively high positions in the royal court, they wished to use their power to eradicate social ills and make their country a better place. Down-ranked, traumatized, and frustrated, they decided to follow in Tao's footsteps and return to *tianyuan*, or the woods and mountains. In a letter to his younger brother Su Zhe, Su Dongpo, in his late years, reflected on his entire life, saying that he was not aware of his mistake that he was entangled in politics half of his life and that he wanted to follow Tao's example in the rest of his life. A real hero, Xin Qiji served in the military for most of his life, helping the emperor defeat enemies and take over cities that were fallen. He realized in his old age that life is but a fleeting moment and no glory stays, so the most pleasant things for him to do were to get drunk, go on outings, and go to sleep. What Xin did after this realization was to tend bamboos, mountains, and waters instead of tending state affairs, or to trade his 10,000-word-long war strategy monograph for books about growing trees. Finally, he discovered or regained his true self and soothed his soul in a simple life at the *tianyuan*.

Perhaps there originally exists a natural realm, or a pure, undisturbed land of serenity, in the depth of the souls of people like Su Dongpo and Xin Qiji. Only after they have experienced what the world has to offer, for better or for worse, can this land reveal itself from within their humanity. In this case, the poets, enlightened now, may be able to produce what Tao is capable of producing. In a sense, their poetry can be viewed as the "difference" of Tao's specter.

In the traditional farming society, Chinese scholars have much leeway in choosing Nature or institutions, the *tianyuan* or the earthly world. In man's living space, there is enough room for Nature. Even in the heyday of agricultural society, which coincides with the Tang and Song dynasties, the conflict between Nature and man, if any, is not as tense as in the industrial age. At that time, there exists *tianyuan*, a buffer zone, or a dwelling where man's soul can safely reside. This dwelling for souls is kept intact in poetry and prose of that day. Since the industrial age, Nature has been shrinking, and the way back to Nature is almost blocked. Now that China is bidding farewell to its traditional farming society and moving forward in the direction of modernization and industrialization, the specter of Tao is faced with an unprecedented threat.

5.3 Can Man Resort to the Specter?

The "specter" theory of Derrida should be viewed as a subtle, significant metaphor rather a theory or discipline in the real sense of the word, since Derrida resorts to the specter to solve the issue of "spirit." Derrida tries to distinguish the "specter" from "spirit" and succeeds in doing so to some extent. He argues that the specter is like, but not identical to, spirit. Derrida does not make "much ado about nothing," nor does he make something complex easy. His purpose is to reveal the dissemination and effects of spirit. In a sense, the specter in the Derridean sense can be called "the spirit within or beyond spirit" and his theory as such "post-phenomenology of spirit" or "deconstructionist phenomenology of spirit."

In the works of Plato, Aristotle, and their contemporaries of the axial age in Habermas's term, spirit is considered merely as an "idea." It is a transcendental, shared set form governing everything as opposed to the nature, law, order, and logic of the phenomenological world. All things of the world are but representations or imitations of nature, law, order, and logic as paradigms or maxims. In Plata and Aristotle, spirit and nature are in conflict. Informed by such rationality in the late nineteenth and early twentieth centuries, the Western industrial civilization reached its heyday, and at the same time, some deficiencies of this civilization appeared. The deficiencies and crises find the most visible expression in these two aspects: the traumatization of Nature and the worsening barrenness of man's inner world, making man have a sense of the "doomsday."

Faced with the deterioration of nature and of spirit, many thinkers let out warning or even desperate voices. In a paper on the Renaissance, James Joyce (1832–1941) says, "[I]f we take the European Renaissance as a point of division, … Should we then conclude that present day materialism, which descends in a direct line from the Renaissance, atrophies the spiritual faculties of man, impedes his development, blunts his keenness?" (187) The consequence, according to Joyce, is that man's victories such as their conquering of the space (or they think they have), the earth, diseases, and follies will be reduced to a teardrop in man's spirituality. Heidegger believes that the nature of the new era is epitomized by the disappearance of God and deities, and therefore, there has come the uprooting of man and loss of his spiritual home. Albert Schweitzer holds a similar view, stating that "The disastrous feature of our civilization is that it is far more developed materially than spiritually. Its balance is disturbed. …Over our progress in knowledge and power we have arrived at a defective conception of civilization itself" ("PC" 86). Ludwig von Bertalanffy (1901–1972) is more straightforward, declaring that "It may be expressed in one brief sentence: We have conquered the world, but somewhere on the way, we seem to have lost our soul" (13).

Confronted with the dual crises of nature and spirit, the West has begun to reflect upon its point of departure and course of development. As part of the reflection, the redefining of "spirit" is particularly worth noting.

The philosophy of life that Wilhelm Dilthey, Henri Bergson, and Georg Simmel represent is, among others, an attempt to challenge the outmoded conception of reason and the deprivation of man's spiritual richness it has partly caused. In Dilthey, one of the major missions of the philosophy of life is to reflect on man's spiritual and cultural activities. He does not agree on Hegel's conception of spirit as abstract idea, nor does he agree with Kant's epistemology in that the net of reason cannot govern man's rich and dynamic spiritual life. The core of the philosophy of life is that the spiritual life is considered as an organic, dynamic, incessant creative process, in which, in addition to reason, life instincts, impulses, intuition, etc., should get involved. The philosophy of life holds maintains that spirit should not be treated merely a rational instrument to deal with the external world; spirit is the value and purpose of life in its own right. To a large extent, the philosophy of life expands the intension of spirit and includes in its scope issues that are beyond reason. This trend is echoed by psychoanalysis of Freud, Jung, and Fromm. For them, the spiritual domain of humanity contains not only the individual's unconsciousness but also man's collective unconsciousness, meaning that hidden in the depth of an individual's soul, there is accumulated culture of the nation. In this sense, spirit is the result of the evolution and accumulation of man's survival wisdom in a certain context.

Informed by Dilthey's philosophy of life and Freud's psychoanalysis, and on the basis of anthropology, Max Scheler establishes his phenomenology of spirit (he himself terms it "the phenomenology of emotions"). In Scheler, the value judgment of spirit is also that of emotions. He suggests that *vernunlft* (reason) should be replaced by the more comprehensive word *geist* (spirit), which contains feelings and emotions such as virtue, love, regret, and fear. According to Scheler, spirit is free, independent, "*weltoffen*" (open to the world); its essence is "no longer subject to its drives and environment" (Scheler "MPN" 52). However, Scheler believes that spirit, as higher form of being of the person, is always feeble because it has no energy of its own, and for this reason, it has to rely on its drives for energy.

If the genealogy of the philosophy of spirit has to be drawn, it should be Dilthey's and Bergson's philosophy of life → Freud's and Jung's psychoanalysis → Scheler's phenomenology of spirit → Derrida's theory of specters.

What Derrida's theory of specters' deconstructs is, above all, the intension of "spirit" in order to go beyond the dualism of matter and spirit. His thought, which is labeled "post-modern," echoes classical Chinese philosophy because in classical Chinese philosophy *jingshen* (spirit) is more often than not referred to as *jing* (essence) or *shen* (spirit), which is very close to "specter" in the Derridean sense of the word.

The word *jingshen* appeared in the fourth century, which coincides with the Western "axial age." It is found in *Zhuangzi*. However, the concepts *jing* and *shen* already existed before *Zhuangzi* was written. In the Chinese tradition, the intension or significance of *jingshen* is usually interpreted as follows:

> *Jingshen* is generated out of Dao. In *Liezi*, it is stated that "*Jingshen* is generated by Heaven and the form by the Earth." In this sense, *jingshen*, or spirit for short, opposed to the form, is mysterious thingless substance of the cosmos; it is completely a metaphysical being.
>
> Pure spirit flows in all directions and reaches all places. It rises up to Heaven above and spreads all over the earth below, transforming and nurturing everything in the world without leaving a single trace. It functions in the same way as the heaven and the earth. (Zhuangzi 251)

"Spirit is the origin of lives" according to *Huangdi Neijing*, or *The Yellow Emperor's Classic of Internal Medicine*. Similarly, Lao Tzu and Zhuangzi believe that Dao contains *jing* (essence), which, in turn, generates and nurtures all things.

That is to say, spirit has unlimited reproducibility and spiritual activities are also a form of production.

For Zhuangzi, the root of Heaven and the Earth is the same as the essence of morality and the law of Heaven is the ideal that man aspires to or follows. The so-called "true man", the "perfect man", the "sagely man", or the "godly man" shares the virtue of *yin* when he is still and shares the movement of *yang* when he is active" (Zhuangzi 249). A man as such "neither contemplates nor premeditates" (249). "His mind is pure and simple; his spirit is staunch and tireless" (249).

Spirit is the concentration of the "vital force" and "nimbus" of the universe. After the body dies, the form returns to the tangible part of the vital force and the spirit returns to the intangible part. This means that life returns to its original state of being. It is worth noting that the spirit a true man or a perfect man has does not vanish as the body ceases to be; instead, it can remain as pure and active as ever. That is to say, it can operate independent of the body and benefit the living.

The above-mentioned interpretations indicate that in classical Chinese philosophy, man's spirit is one with Nature; it is a metaphysical being, a flowing, incessant, continuous vital substance of life and also a component of human nature. In his criticism of the Western theories about spirit, Scheler, to a large degree, identifies with Oriental cultural and spiritual legacy. He believes that in the wisdom of China, Japan, and India, some highest principles regarding existence and life have found their way into Western philosophy. Unfortunately, since one hundred years ago, classical Chinese philosophy has been often tested by the reagent of materialism and idealism and filtered by "scientific thinking." Consequently, the ancient oriental wisdom has been deconstructed by the short-termist modern Western "rationality."

What can we see in classical Chinese philosophy of "spirit" from the perspective of the specter in the Derridean sense?

Zhuangzi says the following:

> A wholesome physical form and sufficient vital energy will enable him to relate with nature. Heaven and the earth give birth to everything in the world. The combination of the physical form and the vital energy results in things; the separation of the physical form the vital energy marks the beginning of new things. The perfect state of the physical form and the vital energy is called the "potential for transformation." A perfection of the perfect state of the vital energy will in turn facilitate the transformation of nature. (299)

The philosopher Wang Fuzhi (1619–1692) interprets this statement as follows:

> The birth of man means that Heaven concentrates qi to form the form which contains spirit. The death of man means that the form returns to the intangible part of vital force, and spirit returns to the intangible part of vital force, thus becoming one with what generates life, or the Beginning. This shows that the bright and the dark are also one. If both the form and spirit can be kept, the pure vital force of Heaven and the Earth can be concentrated in bodies. … Even if the body disappears, spirit remains intact, and continues to benefit all lives. Therefore, it follow's nature's transformation, remains eternal, and omnipresent. (155–156)

One may ask that is there any spirit independent of the body? *Lao Tzu* offers an answer: "When one dies one is not lost, there is no other longevity" (69). Wang Bi's explanation is that the body deceases, but Tao still remains.

Admittedly, no one is certain whether these hypotheses can be proved by science, nor does anyone know for sure whether Jung's hypothesis of collective unconsciousness can be proved by science. Lao Tzu's assertion and Jung's hypothesis follow the same logic in that Jung also believes that the individual's spiritual activities do not stop as the body dies; instead, its spirit remains in the world and finds its way to the collective unconsciousness through a certain channel such as biological or cultural inheritance. Jung explains:

> These ages not only formed an hypothesis about the world system of the spirit, but they assumed without question that this system was a being with a will and consciousness – was even a person – and they called this being God, the quintessence of reality. He was for them the most real of beings, the first cause, through whom alone the soul could be understood. There is psychological justification for this supposition, for it is only appropriate to call divine an almost immortal being whose experience, compared to that of man, is nearly eternal. (Jung "SWJ" 27)

For Jung, the individual's spirit belongs to the world system of the spirit. The former relies on the latter for nutrition and, at the same time, infuses new life and vitality into the latter, thus guaranteeing the operation of the whole world. This is reminiscent of Zhuangzi's assertion about spirit. Coincidently, Jung repeatedly refers to the original meanings of "spirit" and "soul" in English, Gothic, classical German, Greek, Latin, classical Slovic, and Arabic. He points out that they are associated with "wind," "breath," "exhaling," "vitality," etc. The description of spirit as such is similar to classical Chinese philosophy of *qi*, or vital force, since, as mentioned earlier, spirit in the Taoist tradition is generated out of *qi*.

The deterioration of Nature aside, spiritual decline is more severe than ever since the commencement of the industrial age. As a newly emerging modern country, China, due to its unique social structure, history, fundamental realities, etc., has more ills of modernization than modernized Western countries. The worst of the ills finds expression in spirit and Nature. "Deprived of its flesh, blood, and soul, Chinese culture has been reduced to a lifeless skeleton" (Hsu 130). Nature is no better because China's gravest crisis of all is that it is almost inhabitable due to the pollution of most rivers, desertization, and soil erosion. Someday, we may have no living place, no drinking water, and no fresh air.

Can people resort to specters to help modern society out of the impasse?

As discussed earlier, Derrida's *The Specter of Karl Marx* comes as an attempt at salvation in the sense of evoking Marxism. "The specter is not only the carnal apparition of the spirit, its phenomenal body, its fallen and guilty body, it is also the impatient and nostalgic waiting for a redemption" (Derrida 170). The specter has invincible power, since it never dies and "remains always to come and to come-back" (123). After Heidegger's attempt to save the times by poetry, Derrida rests his hope of salvation on the specter.

After Tao's second death as already mentioned, he has become a real specter. Even forced out of its original context and social realities, his specter still remains and wanders. Now, it is no longer subject to the constraint of time and space, or to realities, texts, interpreters, followers, imitators, critics, etc. Compared with the thought, theory, or notion that is captured in one's complete works, analects, or notes, the specter is more subtle, delicate, and penetrating.

A mission of literary criticism is to see what is invisible, or to see the feeble light in the depth of the darkness. In other words, the specter is to be evoked.

Freud defines the nature of literature as a dream, which is akin to the wondering soul. He believes that there is no exact correspondence between a dream, which is a riddle, and its interpretation, which is the answer. Since there is no complete, definite, and fixed relationship between a specter and its carnality as Derrida observes, literary critics are expected to find and evoke the specter. About the modernization in Europe, Jung calls for the restoration of the spiritual bond between man and Nature. He emphasizes:

> Therein lies the social significance of art: it is constantly at work educating the spirit of the age, conjuring up the forms in which the age is most lacking. The unsatisfied yearning of the artist reaches back to the primordial image in the unconscious which is best fitted to compensate the inadequacy and one-sidedness of the present. (Jung "SWJ" 228).

Is the primordial image in the unconscious not a specter in the Derridean sense? The answer is positive. However, how can people resort to specters?

People are expected to change their attitudes toward, and deepen their understanding of, specters so that they may be able to correct their preconception that whatever is labeled as "superstitious" or "supernatural" or "traditional" can be eradicated by man's "invincible power" of which they are so proud. The part of tradition that belongs to the spiritual and cultural domain cannot be uprooted that easily, if at all. The redemptive specters hidden in the depth of human thought and feelings still wander in the woods, on the mountains, in the moonlight, breezes, clouds, mists, and what not, offsetting the overwhelming earthly world that is depriving its habitants of their spirit. It is worth emphasizing that these specters should be conjured up in this new context in order to rebalance the *yin* and the *yang* and to benefit the earth's ecological system by changing how people think and act in the first place.

5.4 The Specter of Tao and the Plights of Contemporary Human Existence

Both Rousseau, a critic of the eighteenth-century Europe, and Thoreau, a critic of the nineteenth-century American industrial society, are in the field of modernity; therefore, their relevance to contemporary society is indisputable. Is Tao, a local farmer poet of about 1600 years ago, eligible for dialogues on the reorientation of contemporary human existence?

To answer this question, it is necessary to revisit Heidegger's observation on poetry and poets. "Human existence is 'poetic' in its ground. … poetry is a founding—through the naming of gods and of the essence of things" (Heidegger 60). This statement highlights the role of poetry. Heidegger believes that "Thoughts of the communal spirit are poetic" (115). The reason is that "Before the thoughts of the communal spirit concern any real things that are, they touch upon the very reality of the real" (114). Heidegger believes that what makes poetic salvation possible is that the poet's soul can be activated, saying, "Inspired by the poetic spirit, the poet's soul is animated because he names the poetic ground of what is real, and first brings it to its essence by pointing out the very reality of it"(115). Liu Xiaofeng's interpretation of this statement of Heidegger is as follows:

> When divinity is gone, or when people are wooing fame and fortune, comfort and pleasure, a real poet would search even beyond national and cultural boundaries for man's lost soul and the trace of gods' hidings. Poetry is supposed to be man's nature, and poets's examples of being man. ... With their innocence, passion, and affection, the poet can let out a hidden call to the world to listen to his poetry and to enable them to realize for the first time what their hometown is. (128–129)

In other words, a real poet should be a prophet who is even able to change the land of historical mist into a poetic land. A real poet must also be a philosopher, and a philosopher would have abundant poetry in their life and contemplation. Tao Yuanming, as a local *tianyuan* poet, is exactly such a poet and, at the same time, a specter in the Derridean sense. For this reason, he is even closer to us when today's world is caught in historical mist and uncertainty. He may help people today, who are wandering in the barren spiritual wilderness, to reorient themselves, to clean the sky and the land, and to search for the lost spirituality of man. If human history is a river, then poetry and philosophy would be an unfathomable abyss that does not flow the way the river does, but they, centered around the issue of "Heaven, the earth, gods, and man," always nourish man's heart and mind.

The Tao's literary world and the tradition he represents are explored in order to evoke an innocent, pure, and simple soul, a soul that enables man to rediscover Nature, to live harmoniously with Nature, and to integrate him with Nature when society is materialized and commercialized, when Nature is overburdened, if not exhausted, and when man is deprived of his soul or spirituality.

Confronted with the multiple crises in the twenty-first century, man should include Tao's poetry on the list of the world's sacred scriptures to help him find the way out. Besides, all texts are in the circle of interpretation. Even sages and holy scriptures cannot escape that circle. One does not expect to see, nor will one see a complete halt of the interpretation of Tao simply due to one's own interpretation. In fact, each and every attempt at reinterpretation is a response to the practical issues relevant to the times also a sort of reappearance of Tao's specter. If people today are willing to look back beyond the moment, they may find shimmering spiritual light of Tao's in the darkness of the historical abyss, revealing some omens and crises of the apparently glorious, dazzling, and alluring world.

5.4.1 The Perception of the Enlightened: The Cage and the Disciplined Society

With as sense of relief, Tao Yuanming confesses in his poem that he returned and regained Nature after he gave up his official life in a cage (Tao 53). A key image in

Tao's poetry, the cage is a metaphor for the freedom-depriving, dreadful, unbearable social existence. Other similar images include "the net of the world," "the tight net," and "the big net" as in "The dark rule over the people is like the net for the fish, or like terrible snare for the birds" (Tao 237).

The image of the cage is not created by Tao Yuanming, nor is it used in this sense exclusively by him. In fact, it is a common name for their life experiences in society or for society itself as perceived by many poets and men of letters, past and present.

Zhuangzi says, "The marsh pheasant has to walk ten steps to find a peck of food and a hundred steps for a peck of drink, but it does not want to be raised in a cage" (45).

Some sages even in primitive times know what it means to be "caged" partly because of their closeness to Nature. It is recorded that Yao, a legendary sagely emperor, wanted to hand over his throne to Xu You, but the latter, on hearing this news, left quickly lest he would be entangled in a cage. Similarly, Shun, another legendary sagely emperor, intended to crown Shan Juan, but the latter refused, saying that he was a man of the field and Nature and saw no reason why he would risk being caged by that crown. Unfortunately, as civilization develops, human beings are being deprived of their innocence and simplicity, and as a result, they end up being caged by the civilization they themselves have created. Emperor Shihuang of the Qin Dynasty united the whole China then, thus creating a tight cage of dictatorship and feudalism. In the Tang Dynasty, Emperor Taizong issued a royal decree that the *keju* system or the imperial examination system be implemented, so that all talents would be in the cage as the emperor himself confided. Since then, the cage of talents remained a vanity fair and an arena for most men of letters in imperial China.

Human civilization sustains humanity, but, at the same time, it cages humanity with norms and institutions. This paradox has been baffling many people even of the broadest vision on the one hand and urges them to explore the possibility of changing their life and society on the other. William Blake (1757–1827) mentions the cage in his poetic line "A Robin Red breast in a Cage/Puts all Heaven in a Rage." According to Berlin's interpretation, "[T]he cage of which he [William Blake] speaks is the Enlightenment, … in which he and persons like him appeared to suffocate all their lives in the second half of the eighteenth century" (Berlin 50).

"Man was born free, and everywhere he is in chains" (Rousseau "DPESC" 45). Rousseau's statement captures the tension between man's inborn naturalness and civilization, which remains unsettled and will not be solved as long as humanity exists.

"The Internationale," a once widely sung anthem in proletariat revolutionary movements in the twentieth century, contains the following lyrics: "[T]he spirit be freed from its prison" and "The world is about to change its foundation." It has been proved that it is possible to change the foundation of the world but extremely difficult to free the spirit from its prison. For instance, after they broke away from the ideological prison of the Czar, the Russians found themselves in the cage of Joseph Stalin (1878–1953), ideological and beyond. The ideological cage built during the "Anti-rightist Movement" and the Cultural Revolution was no less destructive than Emperor Shihua's burning books and burying Confucian scholars alive in 213 and 212 B.C., nor is it less sweeping than the rampant persecution and execution of scholars for writing against political taboos during Emperor Yongzheng's reign (1723–1735).

If Tao was caged because of the System of Dominant Family and institutions whose core is the Confucian "rectification of names," and the cage was wooden, then in an industrial age of plenty, the cage is iron. Max Weber (1864–1920) says, "The technical and economic conditions of machine production which today determine the lives of all the individuals who are born into this mechanism, not only those directly concerned with economic acquisition, with irresistible force" (123).[3] What Weber finds is that the care for external goods should only lie on the shoulders of the "saint like a light cloak, which can be thrown aside at any moment" (123). However, "Fate decreed that the cloak should become an iron cage. …" For the last stage of this cultural development, it might well be truly said: "Specialists without spirit, sensualists without heart" (Weber 124).

Weber is not pessimistic, but honest and wise in his prediction that this situation will not come to an end "until the last ton of fossilized coal is burnt" (Weber 123). The iron cage, which oppresses the individual's spirit and freedom, has become a symbol of the modern industrial society. Compared with the previous wooden cage, the iron one is much tighter and colder.

What may be the root reason for the transformation of modern society into a cage? With science and technology as the point of departure, Heidegger reveals the nature of science and technology modern society depends on: The rule of *das Gestell* governs and controls all humans in that in a developed modern society, all things are treated as raw materials or resources, and human existence is set on *das Gestell*. Simply put, man has created a comfortable seat and then tied himself tightly to it. It is not that man manipulates technology, but that technology (plus the market) manipulates man without man's being aware of. Consequently, man's naturalness and divinity is lost. This is a situation from which there is no escape. "It is a trap that the West has designed for itself against the divine providence, though people enjoy themselves in it and think that is supposed to be the way" (Zhao 72). Albert Einstein states that he has never looked upon ease and happiness as ends in themselves, because for him, this is "the ideal of a pigsty." Like it or not, modern science and technology has changed ours into a comfortable, convenient, and functionalized world, which, in fact, is an invisible super cage overriding all humans.

Michel Foucault (1926–1984) focuses more of his criticism on social mechanisms and the whole human civilization. The intensity and depth of his criticism are remarkable. He believes that what modern social system models on is nothing but the modern prison management. Ancient prisons resorted to physical punishment, but the modern prison of society resorts to power and knowledge

---

[3]Weber (2001).

in taming its prisoners. In the eyes of Foucault, modern society is a giant prison, and in terms of functions, ends, means, and nature, there is no essential difference between a prison and a factory or military barracks. A society as such is called by Foucault a "disciplined society," a key of word of his 1957 book *Discipline and Punish: The Birth of the Prison*. Foucault emphasizes that there is a complete set of discipline techniques with which everyone in society will be developed into a tamed and useful individual. These techniques include a series of manageable means such as monitoring, inspection, quantification, assessment, verdict, and punishment. Foucault points out that modern social governance system is nothing but an imitation of the modern prison system. In his eyes, all institutions and agencies as important components of modern society such as military barracks, hospitals, factories, and schools operate the same way. In this sense, the whole society is a big prison and every individual is a subject to be monitored and transformed. A real prison tends to have harsher disciplines, but the prison of society has wider scope of governance.

Foucault finds that disciplinary institutions are everywhere in modern society; therefore, there has emerged the disciplinary pyramid or archipelagoes. For him, "The French society is a society of prison covered by the net of prison systems" (Liu B.C. 297). Foucault's statement may sound incredible, but nowadays, surveillance cameras are installed everywhere: highways, railway and bus stations, hotels, supermarkets, discos, cybercafe, restaurants, classrooms, and even your room living rooms and bedrooms. One's activities can be recorded, stored, and documented and can be retrieved. Therefore, one may feel that one is imprisoned.

There is tension between the disciplining and the disciplined. There is a strong voice that for the benefits of the vast majority of people, the party, the government, and their leaders should be put in a cage in the first place.

In fact, in the super prison of modern society, no one is completely free. As Jeremy Bentham (1748–1832), the designer of the Panopticon, says, even the highest watchman of the prison is also watched. It is reported that Barack Obama complains about his "being in a bubble" and "followed practically everywhere by staff, Secret Service agents and the media." He says, "It's very hard to escape. Every move you make ... and over time, you know, what happens is that you feel like you're not able to just have a spontaneous conversation with folks. And that's a loss. That's a big loss."[4] Obama seems to be honest and wise in that he knows what his situation is. In fact, his bubble is not vastly different from Tao's cage.

Is it possible to escape from the cage and return to freedom and Nature? Jin Yuelin's answer is as follows:

> One is liable to agree with Rousseau that man is everywhere in chains whether or not there ever was once a state of nature in which he was free. Knowledge by itself is harmonious, but purposes often conflict with one another not merely between states or races

[4]Obama bemoans "bubble" of security http://www.belfasttelegraph.co.uk/news/world-news/obama-bemoans-bubble-of-scrutiny-28586759.html. Accessed on December 20, 2015.

> or different men but also in single individual himself. An individual with conflicting purposes is the enclosed battle ground for spiritual struggle, and although stone wells do not make a prison, nor mere object nature any obstacle to his yearnings, he is yet his own prisoner. The more power one acquires, the more one may be enslaved. ("CWJ" 179)

Undoubtedly, modern people have gained much more power than their ancestors, which finds expression not only in man's control of Nature by virtue of science and technology, but also in man's control of man by virtue of advanced management mechanisms. According to Jin, the only escape from the self-imprisonment is "self-aloofness," at least mentally if not both physically and mentally. Such self-aloofness is what Tao Yuanming calls "the remoteness of the heart makes the earthly world a retreat." This natural, oriental, Daoist escape may be a way, if not the way, for modern people to "return."

Obviously, the way to "escape" has to be chosen by humanity themselves. The question is, would there be any modern paradigms in the nature of Tao's return?

5.4.2 Return: Regressivism and Progressivism

In Chinese culture and literature, the name of Tao Yuanming is always associated with "return." His poetry and prose contain many images of return, such as returning birds, returning men, returning to the field, a heart to return, returning to *the tianyuan*, and returning to emptiness and nothingness. It is claimed that Tao's literary status can be established even by a single piece of his prose, namely "Homeward ho!" Qian Zhongshu (1910–1988) also thinks that this piece is the best written of all the writings of the Wei and Jin dynasties. In Chinese poetry, Tao Yuanming is sometimes compared to a cuckoo, an emperor-reincarnated bird believed to be yearning to return to its home. "Return" is considered as a symbol of Tao's cultural legacy.

Judging from his political career, Tao Yuanming seems to be a born idler, though he has poetic lines about such sublime figures as Kuafu the sun chaser and Xingtian the relentlessly fighting headless warrior. "Water flows downward but man wants to move upward." The better part of the Chinese proverb captures the naturalness of Nature, and the latter part captures human nature and the reason for social mobility. It seems that Tao's decision to move downward conforms to the law of Nature though it runs counter to human's social desire. It shows Tao's identity with the Daoist wisdom:

> The highest good is like that of water.
> The goodness of is that it benefits the ten thousand creatures;
> Yet itself does not scramble,
> But is content with the places that all men disdain.
> It is this makes water so near to the Way. (17)

Lao Tzu says, "How did the great rivers and seas get their kingship over the hundred lesser streams? Through the merit of being lower than they; that was how they got their kingship" (141). Tao Yuanming practices this in his life. His moving downward is out of his identity with Nature heart and soul. Tao describes the freedom and naturalness of things this way:

> Birds in the cage would long for wooded hills;
> Fish in the pond would yearn for flowing rills.
> So I reclaim the land in southern fields
> To suit my bent for reaping farmland yields. (53)

Having their respective comfort zones in the ecology, all organisms, including man, are instinctually attached to their habitats, surroundings, or hometowns. In fact, there is a causal relation between the cage and "return." Since the madding world is a cage, returning to the *tianyuan* is to break away from that cage. Tao expresses his envy of birds and fish again, saying, "How I admire the birds that soar and fly! How I esteem the fish that swim nearby!" (47) After he has decided to return, he writes with excitement: "Homeward ho! Why not return now that my fields will go into weeds?" (245) His reflections of his previous life are as follows:

> There is no need for me to lament by myself.
> I have realized that there is no remedy for the past,
> But there is still a future lying ahead.
> It is true that I was not far astray in my way,
> Realizing that I have turned from wrong to right. (245)

As mentioned earlier, returning is returning to the origin or Nature.

It is safe to say that Tao's poetic return and the philosophical return in the reflections on modernity in the twentieth century are of isomorphic heterogeneity. Henry Thoreau, Friedrich Nietzsche, Max Scheler, Georg Simmel, Leo Strauss, Martin Heidegger, J. Derrida, Michel Foucault, etc., are unexceptionally people who have "returned" after having gone astray.

Nietzsche, for one, is determined to "revaluate all values" of modern society, stating:

> Modern society is no "society," no "body," but a sick conglomerate of chandalas – a society that no longer has the strength to excrete. To what extent sickliness, owing to the symbiosis of centuries, goes much deeper: modern virtue, modern spirituality, our science as forms of sickness. (31–32)

This statement is full of disappointment of modern society. However, looking backward, Nietzsche, who becomes affectionate immediately, finds true men, a nature before the emergence of human civilization, and deities, thus having sublime satisfactions. Indisputably, the philosophy of Nietzsche is characterized by a strong poetic urge to "return."

As mentioned earlier, Heidegger, a soul mate of Tao's, is also a philosopher of "return." In his eyes, the history of Western thought since Plato and Aristotle is but a history of philosophical mistakes, the most striking of which are man's oblivion of existence, excessive dependence on external things, clinging to rationalism, and indifference to nature; therefore, "trivial techniques are prioritized over the great Way" and "man, toiling and moiling in this acquisitive world, is falling" (Zhao 70–73). In a sense, Heidegger's existential phenomenology is an attempt to find a way back home, back to the true, or back to the origin, for the West. For him, a life close to the origin is close to nature, which is similar to Lao Tzu's "returning to the state of infancy" (Lao Tzu 59) or Tao's "nature regained." Heidegger's philosophy eventually returns to Oriental poetic meditation as a way of salvation.

The image of the road is of special significance in Heidegger's philosophy, as can be seen from the titles of some of his works. Essentially, his philosophy runs counter to social progress. For Heidegger, "return" is an important feature of thought since thought itself contains something mysterious, and in some cases, only a path of return leads us forward. More like a trail in the woods less traveled by than a highway, this path of return can lead people to the simple and the true. This image also abounds in Tao's poetry. For instance, Tao writes the following:

Although I yearn to go home to the south,
The lengthy distance makes me shut my mouth.
As contact is cut off by hills and rills,
I write this poem to clarify my wills. (15)

For Tao, no road is too difficult for a heart to return. He also describes the road in the Peach-blossom Springs, showing his soothing detachment from the madding crowd. Whether it be Heidegger's road, or Tao's Peach-blossom Springs, on the road of turn, there are not many souls, if at all. People today are racing forward along the road of modernization by train, car, ship, or plane. One cannot help asking, are they getting any closer to happiness and perfection? That may be just wishful thinking. People today are supposed to keep in mind the admonishment from Heidegger: Man seeks after progress and seems to approach his goal, but when he sees it, it becomes further away.

In his "Going, going" composed in 1977, Philip Larkin (1922–1985) laments what is gone in England, expressing a sense of loss and nostalgia.

And that will be England gone,
The shadows, the meadows, the lanes,
The guildhalls, the carved choirs.
There'll be books; it will linger on
In galleries; but all that remains
For us will be concrete and tyres.

What was true of England three decades ago is also true of China today, because in China, rapidly disappearing are forests, grasslands, streams, wetlands, traditional village and township dwellings, traditional urban landscape and customs, religious rituals, and what not. It is very likely that the past will only be preserved as images and documents in libraries and museums, if not in the collective unconscious of the nation. What "concrete and tires" symbolize are the engulfing concrete jungle and express ways and the overwhelming auto industry. There is little natural vitality in the concrete jungle as there is only slim chance of breathing fresh air in a society suffering automobile dependency syndromes. In the twenty-first century, most of the Chinese end up being enslaved by their concrete dwellings and cars on mortgage. In a sense, people trade their otherwise simple and pure soul for concrete and tires and then their souls become colder and hardened. Obviously, types have replaced much of human's natural movement. It seems that man is just tied onto a high-speed conveyor belt on the expressway, or confined to a concrete cage, thus further distorting his inner nature.

The word "progress" is absolutely a positive word in any Chinese dictionary. In some influential coursebooks of literary history, "progress" is the prerequisite with which a writer or a work is evaluated. Unexpectedly, after thousands of years of social progress, humans realize that "social progress" has many side effects, politically, economically, morally, ecologically, etc., making sustainability extremely difficult. Paradoxically, whether those whose minds are set on progress can make progress as they wish has to be decided by a "regressive" rethinking of "progress."

John Bury (1861–1927), a British historian and progressivist, offers a good genealogical review of the idea of progress. He notes that progressivism, which emerged in the late sixteenth and early seventeenth centuries, is a product of the Enlightenment age. Before that, regressivism had been dominating history studies. For instance, many great thinkers incline toward "regression" in terms of view of history such as Plato and Aristotle who believe that there once existed a "golden age" of purity, simplicity, and naturalness and that the later development of human society just diverged from that road. For them, history has been witnessing man's fall again and again. Such an idea is similar to what is implicitly expressed in the Bible. Of course, the list of regressivists should also include Confucius, Mencius, Lao Tzu, and Zhuangzi, for whom the "golden age" was the long span of time covering the Xia, Shang, and Zhou dynasties. Committed to inheriting legacy, Confucius believes that an, if not the, ideal personality was Duke Zhou, 500 years his senior. Mencius argues that the law of previous kings should be followed; he views the legendary Emperors Yao and Shun as perfect examples of sageliness. Lao Tzu and Zhuangzi are even more inclined toward regression, arguing that man should abandon his wit and knowledge, benevolence and righteousness, and treachery and benefits, or, in other words, all forms and representations of excessive civilization, so as to return to the state of simplicity, infancy, and blissful chaos.

Bury points out that it is Francis Bacon and Descartes who have destroyed the regressive view of history. Bacon advocates that "knowledge is power" and that technology is an instrument for satisfying human needs; Descartes declares that man has an independent position in relation to nature and that man's rationality, which is supreme, is the source of power for knowing and controlling nature. Informed by Bacon and Descartes, "the idea of progress could germinate" (Bury 36). Since then, the ideas of "anthropocentrism," rationalism, the opposition of man and nature, technology as an instrument for utilizing nature, consumption for happiness, etc., have become core notions of progressivism. In his 1924's monograph *Philosophy of Life*, Fung Yu-lan lists a number of progressivists, including Descartes, Bacon, and Johann Fichte. He summarizes progressivism as the idea that "Man, as opposed to Nature, can triumph over Nature with his own intelligence" (Fung "II" 132). Fung seems to have captured the core of progressivism.

After three hundred years of social progress, man has come to a sudden realization that progress is not a complete blessing: What came along with it are curses, catastrophes, omens, and nightmares.

The two world wars are proof to the distortion of human nature. The large-scale gas chambers of the Auschwitz Concentration Camp and the nuclear bomb exploded over Hiroshima were definitely highly efficient killing instruments. One cannot see obvious progress of human nature with the development of science and technology. In recent years, the ever-widening gap between rich and poor, the rising tension between nations, the spreading of terrorism, etc., seem to have justified man's skepticism of progress. At the same time, a sense of happiness does not come along automatically with industrial and technological development or increased consumption. On the contrary, people's life quality and moral standards are believed to be declining.

What sound a bigger alarm for progressivism are the reoccurring ecological catastrophes and the impending ecological crisis. Even air and water is limited, not to mention animal and plant species and mineral reserves. How is that possible to pursue unlimited progress in a limited space with limited resources?

In today's community of social sciences, especially of historical studies, there is growing skepticism about progressivism after its predominance for almost three hundred years.

The Hydra (a monster with nine heads) effect in complexity theory implies that social progress is now in a dilemma, because the solution of one problem causes more problems, thus pushing man into a deeper plight. As society progresses, problems multiply, social governance swells, and risks increase. The assertion of complexity theory is quite pessimistic, arguing "Where double the number of steps must go right in a goal-pursuing context, we double the prospects of some things going wrong" (Rescher 182). It is argued that man's capability of dealing with larger system of complexity will reach its limit someday, and all the artificial systems may be rendered invalid by contingencies, thus causing disastrous consequences. As for countermeasures, what complexity theory can do is to offer some admonishments such as "moderation" with our material desires and compromise with nature's constraints. It also admonishes that man should acknowledge his limitations, the plight of rationality, and the incompleteness of his knowledge, stating that "ignorance is bliss" (Rescher 191). Not surprisingly, complexity theory, which is of strong post-modernity, seems to "return" to the road of Lao Tzu's "regression" philosophy whose core is "simplicity" and "the abandoning of knowledge." It is advisable that in today's world, people, fatigued, should hold fast to what is natural and simply their life accordingly.

There is abundance of images about returning to nature in Tao's poetry. "Returning birds," for one, captures the sense of harmony and naturalness in nature. It reads:

> Returning birds glide in the sky;
> On chilly boughs they stop and lie.
> In the woods they play and tease;
> At night they sleep atop the trees.
> In the early morning breeze;
> Are heard the pleasant songs in speers.
> Hunters, spare your arrows, please!
> Tired birds are hidden in the trees! (61)

In the Chinese tradition, there exists a strong urge to reject civilization progress even from the very beginning of civilization, which finds best expression in the Daoist philosophy of Lao Tzu and Zhuangzi. The primitive society before the civilized one is usually believed to be an ideal society, or a society of great virtue. Here is a vivid account:

> The people, acting in accordance with their natural instincts, weave cloth to get dressed and till the land to get fed. This is called their "uniform integrity". They think and act like, with no partiality against each other. This is called their "natural freedom". Therefore, in ancient times when perfect virtue prevailed, the people walked in self-contentment and did not care to look around. At that time, there were no trails or paths in the mountains and there were no boats or bridges the waters. Everything thrived on earth, with no barriers between each other. Birds and animals could be tethered and led about. People could even climb up to the magpie's nests and peep into them without disturbing them. In ancient times when perfect virtue prevailed, people lived together with birds and animals and mixed with everything in the world. How did they know the distinction between superior men and inferior men? All ignorant, they did not lose their virtue; all desireless, they were in a state of natural simplicity as uncarved timber, which kept intact their inborn nature. (Zhuangzi 135–137)

These passive regressive ideals are seldom adopted by mainstream society since they do not conform to social norms, political correctness, the rulers' benefits, and even the needs of the public. Traditionally, "return" is just an alternative or the last resort for Chinese men of letters with only an exception, that is, Tao Yuanming, who chose to return. Tao confesses, "I would live a simple life cherishing my aspirations than sell myself to the court at a high price" (241). The significance of Tao's choice in a utilitarian society has not been fully understood for a long time. Only when people today reflect on modern society, only when they have a new understanding of the ecological situation, and only when ecology has become a global issue, can Tao's significance be fully revealed. Jean-Franois Lyotan (1924–1998), a post-modern French philosopher, points out that "For me 'ecology' means the discourse of the secluded. ... This discourse is called 'literature', 'art', or 'writing' in general" (Lyotard 135–139). For him, the times has compressed man into an electronic public file, thus partly blocking the road of man's return and shattering Nature's God. As a result, the pastoral has become an elegy, and the *tianyuan* just relics of nostalgia.

5.4.3 A Pure Heart and Cheerful Chanting: Honorable Poverty and the Consumer Society

In his lifetime, Tao Yuanming did not have fortune or nobility. He confesses that "As I was underfed all through the year, I left the countryside for my career" (125).

"The wren that builds a nest in the deep forest occupies only a single branch; the mole that drinks from the river takes only a bellyful" (Zhuangzi 9). This means that minimalism, more often than not, makes a better life. Holding fast to Daoist teachings, Tao wanted to live a self-sufficient, simple, peaceful, and idle life, saying, "There is no need for a spacious hut instead, if it can keep rain off my mat and bed" (71). In fact, in case of natural disasters, poor yield or crop failure or other contingencies, Tao and his family would suffer hunger and cold. His difficulty is described in some of his lines as follows:

> Starved in summer I am in a plight,
> Have no quilts for the winter night. (135)

> I wear the threadbare garments as my dress,
> And often eat wild herbs in times of dress. (143)

> In the jugs there's not a drop of wine;
> On the stove there's not a dish to dine. (141)

> Driven by hunger I seek for food and drink,
> Not knowing where to go however hard I think.
> I walk and walk till I come to a door;
> I tap the door but stammer, plead, implore. (207)

By today's standard, Tao definitely lived deep under the poverty line. Despite this, he did not give up on his aspirations and ideals, confessing the following:

> I would rather stick to my aspirations in poverty
> Than tire myself out in abandoning my aspirations.
> Now that I do not regard a high position as a distinction,
> How can I feel my disgrace when I am dressed in rags?
> If I would come to no good in my official career,
> I am ready to give up my position and live like a recluse.
> I would live a simple life cherishing my aspirations,
> Than sell myself to the court at a high price. (241)

As always, Tao maintains honorable poverty, a virtue highly valued by traditional Chinese scholars. He describes himself this way: "I've stuck to honest poverty sublime but suffered cold and hunger all the time" (123). Despite his impoverishment and occasional starvation, he maintains his dignity and happiness and creates a life of quality, at least spiritually, by getting closer to Nature, getting well along with his fellow villagers, and tilling the land. His anthology is full of cheerful chantings about the beauty and pleasure of such an existence. Some typical descriptions are as follows:

> It's fine today and we relax at ease,
> In pleasant music of the flutes and strings. (105)

> When new songs ring around in cheerful glee,
> We drink the fresh wine to our hearts' content. (105)

> Not a cloud floats in the evening skies;
> Gently blows the spring breeze from sunrise.
> Throughout the night so silent and forlorn,
> A lovely maiden drinks and sings till dawn. (159)

> The Hermits' songs ring loud in my heart
> Although I'm ill in fate to stay apart.
> The age-long sorrow in my heart I bore,
> Long after songs died out and rang no more. (107)

> The old Rong Qiqi, in a girdle crude,
> Played the zither in a cheerful mood.
> Yuan Zisi, in his shoes through wear and tear,
> Raised his voice and sang an ancient air. (143)

Tao views Rong Qiqi and Yuan Zisi, hermits in a time of chaos, as his soul mates. For Tao, "being poor" just means having scant material possessions; it is not a disgrace. A real disgrace lies in one's failure to practice Dao even though one knows it. To this, a remark of Zhuangzi serves as a good footnote:

> In ancient times, the men endowed with [D]ao found joy and pleasure in both favorable and unfavorable situations. Their pleasure did not come from visible disadvantages or advantages. With [D]ao and virtue deeply embedded in mind, they believed that disadvantages would turn to advantages and vice versa in the way that winter would take the place of summer and rain would follow wind. (503)

For the enlightened ones, poverty cannot undermine their pleasure and happiness. Their being poor and honest is called "honorable poverty." Tao writes the following:

> The Confucian teaching rings without doubt:
> It's learning, not poverty, that man cares about.
> The teaching is easier said than done,
> And so I turn to work hard in the sun. (25)

Taoists and Confucianists share the idea that integrity should be maintained even in poverty. Nevertheless, the way of Confucius and that of Lao Tzu are vastly different in that the former is not without pragmatism, as can be seen from Confucianists' wish to be recognized and promoted by officials, while the latter, which is also the Heavenly Way, is characterized by "non-possession," "non-clinging," and "self-effacement." Confucius himself traveled many kingdoms to advocate his teachings and, to this end, had to deal with the royal courts and dignitaries. In contrast, Lao Tzu chose to retreat into a no-man's land after he abandoned his position as the curator of a library, and Zhuangzi spent his lifetime tending a garden of lacquer trees and making straw sandals. Confucianists may keep honorable poverty passively, but Daoists can obtain indifference and aloofness proactively in addition to keeping honorable poverty. It seems that poverty helps Daoists achieve Dao. "They forget everything but possess everything; they show complete indifference but with the best fame. Such people embody the eternal [D]ao which permeates the heaven and the Earth" (Lao Tzu 247–249). This quote highlights the importance of non-possession. For Zhuangzi, "Average possession means happiness; surplus means trouble. This applies to all and it is even more obvious with material acquisition" (535). For Daoists, affluence and fortune are not conducive to the realization of Dao, or to the purity of the mind or the serenity of the heart. Tao Yuanming says, "When such awareness as of myself is lost, why should I care about this thing or what it will cost?" (121) What Tao means is that possessions are useless if one is aware of one's own existence.

In classical Chinese, *qingpin*, or "honorable poverty," is almost always a positive phrase, which is used to describe the shortage of material possessions and more often to describe the character of men of letters. *Qing*, meaning "clear," "innocent," "pure," and "tranquil," is surely positive; *pin*, meaning "absent," "scanty," "insufficient," and "scarce," is not derogatory in Daoist philosophy since "what is most perfect seems to have something missing; … what is most full seems empty" (Lao Tzu 97). Daoism values emptiness, nothingness, tranquility, frugality, and simplicity; it opposes possession, fullness, luxury, and extravagance. This almost adds up to *qingpin*. Tao's honorable poverty, which is more of a choice of free will than of helplessness, embodies the Daoist spirit of simplicity and naturalness (so-of-itself). Undyed silk and uncarved log are metaphors that are frequently used to describe the naturalness of things, or the Heavenly Way. A successful return to simplicity and naturalness is the highest achievement of Daoist practitioners. In this sense, Tao's returning to the tianyuan also means returning to simplicity and naturalness. Therefore, honorable poverty is no longer something that mars man, but something that makes man in terms of returning to Nature. The "simple vista of life" as advocated by Jin Yuelin is exactly about returning to simplicity and naturalness, a state akin to "infancy" in Tao Tzu's term.

Obviously, society has not regressed to the state of infancy. On the contrary, it has grown into an almighty giant, though it is wandering further away from the origin and the true. "Since the ancient practice of simplicity disappeared, the evil practice of hypocrisy has developed" (Tao 235). What Tao criticizes here is the officials' lack of virtues. However, the greed and hypocrisy of modern people, especially those officials and business people that are far more serious than in Tao's day.

In fact, modern people view poverty as a plague that everyone is supposed to avoid. As a result, competition becomes every increasingly fierce, which, in turn, causes the polarization of fortune. It comes as no surprise that there has been greater tension between social strata and between nations. Long gone are the days when peace, serenity, and contentment prevailed. Inner nature aside, some people today are not even serious and sincere about their own natural bodies. On the mass media, there is passionate advertisement about "man-made beauties." The "modification" of a natural female body may involve a series of surgical operations such as silica gel injection, liposuction eyelid doubling, rib removal, and skin removal, the brutality of which can only be matched by the most merciless of physical tortures of old days. Many are willing, if not eager, to receive such plastic surgeries, which would have horrified Zhuangzi who was saddened by the horse being bridled and the ox being led by the nose.

Undoubtedly, Tao's honorable poverty is close to the Daoist spirit of simplicity and naturalness. *Qing* helps to develop a healthy spiritual ecology, and *pin* helps to maintain the natural ecology. It is safe to say that Tao Yuanming is a perfect example for ecological writers. Wang Xianpei, among others, values the ecological significance of Tao's poetry, arguing:

> Amongst ancient great Chinese poets, Tao is the most enlightening in terms of ecological dimensions and practice. … What he pursues is individual spiritual freedom, characterized by his spiritual needs not succumbing to material needs, and tranquility and ease generated out of the harmony between Nature and man. … It is not accurate to summarize Tao's

> thought into his experience of and proof to the proverb 'Enough is as good as a feast, because it is not an issue about sufficiency or want, but about whether or not man should have spiritual pursuits, about the orientation of one's life philosophy, and about the gradation of the individual's and society's spirit. (479–282)

Wang believes that Tao's thought on ecology is an important piece of wisdom for humanity.

However, this wisdom of Tao's seems to have been abandoned in his own country. In terms of consumption, one striking change in Chinese society in recent years is luxury consumption, which sounds unbelievable.

Various luxury products' consumption reports show that luxury consumption in China has been rising rapidly. It is reported that in 2010, the shopping volume of China's international tourists rose to number one in the world, accounting for 17 % of the world's total. It is predicted that by 2020, China's consumption of the world's most famous fashion brands will account for 44 % of the world's total. Luxury product consumption in China has become a major driver for the world's luxury sales.

In a sense, luxury consumption has become a brand new form of "opium" in twenty-first-century China. As an old Chinese proverb goes, "Aspirations wane as indulgences in possessions wax." The Chinese today have money or possession fetishism. Consequently, the more powerful, magnificent, and noble their possessions are, the more feeble, fragile, and empty their personality is. Surprisingly, how come China, with a long tradition of frugality, and as a passionate advocate of frugality as seen from its slogan song "Frugality is our Legacy" during the Cultural Revolution Period, has succumbed to the fashion of luxury consumption, which is inherently something that underpins capitalism?

In classical Chinese, there is no such word as "*xiaofei*" (consumption). Human beings do consume in the sense of having to consume materials and energy to sustain their lives and creature comforts, but such consumption is necessary and legitimate.

In modern society, the nature of consumption has changed in that what governs man's consumption is no longer the ecological law, or spiritual and moral pursuits, but the law of value of capitalism. Capital stimulates man's consumption desires by virtue of high technology in exchange for high returns. To a large degree, consumption is an instrument by which capital develops the market for profits. To some extent, man does not consume because of needs, but consume for the sake of consumption; man does not produce for consumption, but consume for production, thus changing the whole society into a huge machine that produces consumption products and desires. Modern China does not have a strong capitalist tradition, nor does it have a strong immune system for capitalism; therefore, Chinese consumers are much easier targets of the market and much more likely to be fixed tightly onto this machine, thus completely losing themselves.

Jean Baudrillard (1929–2007), a French thinker and strong voice against consumerism, reveals how totalitarian technocracy of the West has established its control over society through consumption. For him, consumerism is the most evil logic in capitalist economics because, ironically, a man's or a country's success and power is defined by his or its consumption. Ours has been reduced to a "consumer society," a "throwaway society," or a "garbage-can society," and in such a society, "the order of production is cynical" (Baudrillard 24). "Consumption, as a new tribal myth, has become the morality of our present world. It is currently destroying the foundations of the human being" (Baudrillard "Foreword"). Baudrillard's idea has found widespread acceptance, but, unfortunately, endorsing one's idea is one thing, and changing one's pattern of consumption is another.

China Ecological Footprint Report 2010 issued by China Council for International Cooperation on Environment and Development says that China's ecological footprint was already twice the biocapacity in 2007 and that China's ecological deficit has been growing annually.[5] The potential risks of overdraw as such are extremely huge, especially for China, which has a large population and low per capita resources.

In fact, the significance of the economic boom of the world has been undermined by ecological and spiritual decline. To some extent, a healthy simple lifestyle has been replaced by corrupt consumerism. However, consumption does not necessarily increase man's sense of happiness. Alan Durning believes that "The relationship between consumption and personal happiness is weak" (23). He cites a piece of argument that "People living in the nineties are on average four-and-a-half times richer than their great-grandparents were at the turn of the century, but they are not four-and-a-half times happier" (23). He agrees on the judgment of a psychologist at Oxford that "The conditions of life which really make a difference to happiness are those covered by three sources—social relations, work and leisure. And the establishment of a satisfying state of affairs in these spheres does not depend much on wealth, either absolute or relative" (Durning 42).

With the above-mentioned criticism of consumerism as a frame of reference, one may identify the sources of Tao's happiness through a single reading of a poem of his. The poem is as follows:

> In spring and autumn there are sunny days,
> When we climb the hills and write new lays.
> If neighbors pass my door, I'll call aloud
> For them to have a sip if wine's allowed.
> In busy seasons, we go to fields again;
> At leisure time, we miss each other then.
> On that occasion, we put on coats and go,
> Talk and laugh while time goes in a flow.
> There's nothing better than this pleasant year;
> In no case shall I leave my fellow here.
> As food and raiment all come from the land,
> I'll work hard to earn a living by my hand. (Tao 73)

In 408, after his former home was burned down in a fire, Tao Yuanming moved to a more remote and inaccessible South Village, where there lived some people of the same mind. In this poem, Tao describes the harmony between him and his neighbors, farming, and leisure, all the three elements of happiness as believed by

[5]Zhang, Ke. China Ecological Footprint Report 2010: Ecological Deficit Growing. http://www.yicai.com/news/2010/11/594510.html.

Durning. Life as such is poor, healthy, simple, and happy. In classical Chinese, *qingpin* is not an economic term; instead, it is morally and spiritually loaded.

It is beyond the expectation of the advocates of the Enlightenment that rationality, when carried to extremes, changes man into economic man or even monetary man. Baudrillard calls on people to maintain critical "anti-discourse" for "the discourse of consumption and its critical undermining" and offers a solution for salvation by reminding people of Chaban–Delmas's famous flight of oratory: "We have to control consumer society by giving it back some soul!" (195) This oratory also explains why Tao's soul needs to be resurrected in this ecological era.

5.4.4 The Tranquil Southern Mount: Leisure and Work Ethic

Tao Yuanming was probably the only great ancient Chinese poet who was engaged in farming for much of his life. Labor, therefore, can be a good point of entry in studies of Tao.

In anthropological and sociological terms, labor is a significant and complex issue for it is interwoven with society, nature, and man's spiritual ecology, especially in modern society. Tao's return and farming, labor and leisure, and their literary representations may inform us in many ways.

There is nothing wrong about these statements: "Labor is lofty," "Labors are holy," "Labor created man," "Labor created the world," and "No labor, no food." Nevertheless, these prepositions are valid only in a certain context. In other words, in some context, labors can be lowly, and labor can be humiliating, create an ugly world, and ruin man as it makes man.

Max Weber believes that sociologically, labor in an industrial age is essentially different from that in an agricultural age. The latter is laborious, loose, flexible, seasonal, individual, or family-based in terms of organization and therefore less restricting and less efficient; the former, due to the application of machines and to the increased efficiency of these machines, labors are fixed to the machines and to specialized and professionalized organizational areas; therefore, they have high psychological strains though less physical output. This form of labor is termed by Weber as "rational capitalistic organization of formally free labor" (34) and is believed to be "the origin" and "central problem" (36–37).

Karl Marx emphasizes that labor of the industrial age is an organized force consisting capital, means of production, and labor force, and this form of labor is the basis for capitalist society. Max and Marx have a consensus on the nature of labor in the industrial age, though from different perspectives: The former is focused on spiritual culture and the latter politico-economic issues.

The Marxist class theory about labor aside, labor is essentially associated with the issue of Nature and man, the real meta-question of humanity as mentioned earlier. Unfortunately, this is not treated as a serious question in most previous labor theories.

The definitions of labor in Chinese dictionaries are usually based on either Hegel's idea that labor is power man possesses in his tools over external nature, or Marx's idea that labor is a process of exchange of matter between man and nature. For instance, labor is defined as "man's purposeful activity to change nature for his own needs by virtue of tools and at the same time to change himself. Man, as a form of natural power, and standing in opposition to natural matter, acts directly or indirectly on natural subjects so that he can possess natural subjects in a useful way" (Feng Q 790). The core of the definition is that man transforms and possesses nature to satisfy his own needs. In this sense, labor is a tool by which man obtains possessions and exploits nature. The way capitalists deal with workers is similar to the way man deals with nature, but the former is believed to be unfair and even evil and the latter legitimate. Such work ethic should be reconsidered in an age of ecological crisis.

In developed and some of the developing countries, labor, in many cases, is no longer for the needs of survival, but for insatiable desires of consumption. Much of the grand-scaled and highly organized labor today such as deforestation, levelling a mountain, and building expressways, great dams, skyscrapers, and amusement parks is for surplus or luxury consumption. Consequently, man's greed is further aroused and social ethics corrupted in the first place; much of the non-renewable resources are depleted, and the ecology is ruined in the second. Scheler thinks that such "labor" is stinky, barbarian, and void of moral fragrance as existent in labor in the pre-industrial age. Encouraged by the modern conception of labor, the polarization of rich and poor is getting increasingly severe, and therefore, global social stability and harmony may not be within reach in the foreseeable future. Worse still, this distorted conception of labor contaminates man's soul, the inner nature as well. Baudrillard laments that in a consumer society, "Just as there is a world hunger problem, so there is now also a worldwide problem of fatigue" (157). Ironically, modern people whose minds are set on consuming for pleasure end up overdrawing themselves, physically and mentally. With the acceleration of modernization, "pains" become increasingly disproportionate to "gains," which means that more efforts have to be made for the same dose of happiness.

Leisure activities and nature have already become expensive commodities. Breezes, sunshine, the azure sky, the starry night, etc., are all calculated as part of the price for resorts, eco-tour destinations, hotels, and restaurants. In today's China, one may have to pay a considerable price, though in an indirect way, for a view of the starry night or a moment in the fresh air in the country. For most people, labor is the prerequisite of leisure. They have got stuck in the vicious circle of "consumption–labor–consumption–labor."

Now, let us get back to Tao and see how he spent his day.

After having resigned from his post, Tao had to till the land in person to survive, as he describes in his line: "As food and raiment all come from the land, I'll work hard to earn a living by my hand" (73). His daily routine may have been like this: "I rise at early dawn to weed and prune, Till, hoe on shoulder, I return with the moon" (55). Tao did not have any ambition economically because "There is no trouble to meet my daily deed; It is against my hope to be obsessed by greed"

(19). Fortunately, he had leisure. He confesses that "When I have ploughed the field and sown the seed, I, now and then, find time to write and read" (175). He reads and writes for pleasure, not practical purposes. He would converse with his neighbors over topics such as tending mulberries and flaxes and enjoy familial bliss on excursions into the wild for the warbling of birds and the vigor of all animated beings. He would not expect his children to go places and rise high, because to live in nature's arms is already lucky enough and a big relief. His heart would be soothed by writing, especially by the evoked imagery like chrysanthemum plucking in his aesthetic experience.

In a word, Tao epitomizes the ideal, if not idealized, life of plowing and reading in traditional Chinese society, a life of labor cushioned by leisure, ease and joy. By today's standards, such is a genuinely low-carbon high-quality lifestyle. In this case, Tao's prominence does not lie in his engagement in physical labor, or his mingling with his neighbors, but in his integration of labor and leisure in a perfect both in his everyday life and in his literary writings. This may be the state of being that may attract Heidegger profoundly, which can be best described by a line of Hölderlin as follows:

> Full of merit, yet poetically, man
> Dwells on this earth. (Heidegger 60)

This oriental-style "ethos" seems to be natural solvent for capitalism. According to Max Weber, it is partly due to such easy, idling ethos that capitalism has not developed from within Chinese society (at least up to Weber's day). No one is certain whether this is a blessing or a curse.

However, for the sake of global ecological security, it is advisable that people should reduce their labor to be "paid" for surplus consumption and have more leisure and idleness ("the most attractive and productive industry" according to Thoreau).

Viewed from today's work ethic, Rousseau, Thoreau, and Tao Yuanming, epitomes of beautiful ecological existence, seem to be "idlers." For Thoreau, idling in nature can bring him the greatest of pleasures, "but men labor under a mistake" because "the laboring man has not leisure for a true integrity day by day." Similarly, Rousseau's own experience proves that true happiness cannot be measured by money.

The Chinese proverb "Books make you rich and leisure makes you immortals" highlights the value of leisure. After having returned from his exile in Hainan, Su Dongpo, who had experienced numerous rises and falls, discovered what life was and decided to be "an idle man," playing harp music, drinking, and watching clouds. In today's China, there are no longer organized campaigns to repress intellectuals, but scholars do not have as much leisure or idleness. Compared with their predecessors, they are in want of inner serenity, since many of them are focused on publication, research project application, title promotion, book signing, working as judges, bosses or shareholders, appearing on TV shows, etc. As a result, they gain everything except learning and leisure. Presumably, ease and idleness are not only associated with personal state of mind, but also associated with the nature of social institutions and the ethos of the times.

The idea of hard work on a daily basis is not always popular, especially among the younger generations. Such changes, not fundamentally though, can be observed in grassroots "freeters" in the sense of young people deliberately choosing not to work. Without ambitions or aspirations, many freeters think leisure and ease are far more important than work ethic or a decent job. With work skills, they are not parasitic, but they refuse to dedicate themselves entirely to their work as do their fathers. What they want is to satisfy their basic everyday needs to have some freedom, freshness, and leisure; therefore, they are playfully labeled as gypsies of the workplace. For them, labor is no longer the first and foremost necessity of life, nor the means to transform the world, nor a holy deed. Instead, the meaning of their lives begins to return to life itself. It seems that Thoreau and Tao dwell in them. Helplessly, on their way returning to the true and the origin of life, freeters are confronted with the iron curtain of the modern industrial age. A post-modern return to nature may be more of a vision than a reality.

However, the freeter phenomenon is a sign of the change of the times, after all. If it is not considered as a sign of the resurrection of Tao's specter, then it should be interpreted as a message that man's aspirations to naturalness and freedom have begun to regerminate in the minds and hearts of the younger generation and they can go through the iron curtain sometimes some way. This is another case of what Lao Tzu says "In [d]ao the only motion is returning" (87).

5.4.5 Chanting in the Field: Tianyuan Poetry and Agricultural Society

Tao Yuanming is the founder of the *tianyuan* school of poetry, and his presentation of the *tianyuan* landscape and the *tianyuan* complex is a major contribution to Chinese literature. In this sense, Tao's poetry remains a significant legacy in Chinese spiritual culture.

It is necessary to refresh our memory about Tao's presentation of the *tianyuan* landscape. The following is a tribute to the *tianyuan*:

> When clouds o'er hilltops disappear,
> The dim horizon soon turns clear.
> From the south mild breezes blow
> And toss the seedlings to and fro. (Tao 39)

Without bodily experiences, this true, fresh, and joyful experience which contains both dynamism and serenity would have been impossible. Tao had profound attachment to the land and farming, which is expressed by the following lines:

> As rural life to me is dear and near,
> How can I stay away all through the year? (51)

> As I am tired of worldly toil and moil,
> I am tightly attached to native soil. (13)

I close my wattled gate and sing aloud,
Content to be away from the madding crowd. (25)

Farming, of course, is an important part of Tao's poetry. A true expression of his worries and hard work associated with farming is as follows:

In poverty I live and till the land,
Working hard with Donglin close at hand.
At sowing in the spring I take great pains,
In constant fears that harvest gives no gains. (101)

Some of Tao's poems are full of joys of a promising good harvest:

When we meet, a few remarks will go:
How fast the hemp and mulberry leaves grow!
The hemp and mulberry are growing fast;
My ground, my garden plot is growing vast. (55)

Sometimes Tao's pure pleasure comes from drinking with friends:

My friends who share my bobby drinking wine
Come to me with liquor sweet and fine.
We sit below the pines on faggot grounds
And get a bit drunk after several rounds. (121)

Not a cloud floats in the evening skies;
Gently blows the spring breeze from sunrise.
Throughout the night so silent and forlorn,
A lovely maiden drinks and sings till dawn. (159)

Liang Qichao, for one, thinks that Tao Yuanming, due to his perfect description of country life, is the avatar of the beauty of country life. In fact, what is depicted in Tao's poetry is true to traditional Chinese agricultural society.

Farming has been a long-established tradition in China, even in remote antiquities of the legendary kings. In his poetry, Tao himself points out:

Who on earth was our saint then?
Houji began to teach the men.
What way of life did Houji teach?
The farm-work was within their reach.
The ancient rulers ploughed the fields
And always cared for yearly yields. (27)

In later dynasties, the emperors would plow the land in springtime ritually to pray for good harvests and to set an example of valuing farming. What is called the five-thousand-year-long Chinese civilization is in fact a five-thousand-year-long agricultural society. The ethos of the Tao's poetry is exactly that of this society.

In agrarian Chinese society where agriculture is paramount, being an official or being a farmer is highly encouraged ways of life. Due to its closeness to nature, such a life and its rises and falls are more likely to have a poetic touch. In this sense, *tianyuan* poetry can be considered as an elevated representation of the spirit of agrarian Chinese society. Glen Love, among others, particularly values the tradition of pastoral, arguing that "The lasting appeal of pastoral is a testament to our instinctive or mythic sense of ourselves as creatures of natural origins, those who must return periodically to the earth for the rootholds of sanity somehow denied us by civilization" (Glen 230). Love even views pastoral as a resource that can guide humans to return to a life of simplicity and naturalness, stating:

> In the pastoral tradition we have a long and familiar heritage in literature which purports to … examine this complexity. … Literary pastoral traditionally posits a natural world, a green world, to which sophisticated urbanites withdraw in search of the lessons of simplicity which only nature can teach. There, amid sylvan groves and meadows and rural characters – idealized images of country existence – the sophisticates attain a critical vision of the good, simple life, a vision which will presumably sustain them as they return at the end to the great world on the horizon. (231)

That pastoral or *tianyuan* is so vital and appealing is associated with its closeness to man's origin and it is a harmonious blending of the natural ecology, the spiritual ecology, and the social ecology. As is known, "eco" in "ecology" is derived from Greek *οἶκος*, meaning "house." "Ecology is the study of 'life at home' with emphasis on 'the totality or pattern of relations between organisms and their environment'" (Odum and Barrett 2005). According to this definition, Tao's tianyuan poetry is also ecological poetry, and epistemologically, ecology is the study of man's home or dwelling.

Since the Enlightenment Movement, the *tianyuan* life has been disappearing due to industrialization and urbanization, which, in turn, seems to have special attraction for China as a latecomer in the industrial world. In China today, a large-scaled government-dominated urbanization campaign in well under way. Over a short span of 30 years, the rate of urbanization in the mainland of China rose to 43.9 % in 2006, compared to 17.92 in 1978; it will rise to 72.9 % in 2050.[6] As a result of this "stormy" campaign, the root of traditional Chinese society has uprooted, and inevitably, the *tianyuan* landscape is destroyed, if not eradicated. Homes on the *tianyuan* are being relocated or simply demolished, sometimes by force, and replaced by either "modernized living quarters," modernized produce manufacturing bases, or industrial development zones.

Unlike Tao's age, this age is characterized by the polarization of agriculture and industry, and by the destruction of the former by the latter. The worst scenario is that rustic villages are deprived of its vitality and potential by industrialization and rendered barren, with only senile villagers and "left-out" children living there. He Xuefeng concludes that the traditional rural civilization has disintegrated, the traditional tianyuan landscape has disappeared, and traditional peasants have been transformed; therefore, the countryside today is no longer able to provide a meaningful system of life to sustain villagers. The peasant workers' way of "returning to the countryside" has been blocked by the heavy concrete and iron walls of industrialization, and they have been rendered "homeless" economically and emotionally. This explains why migrant peasant workers in the city are very unwilling to return to the country.

---

[6]See http://www.chinadaily.com.cn/dfpd/jingji.

In fact, peasant workers themselves are the eyewitnesses to this dramatic change of the countryside. Some peasant worker poets capture this devastating transformation in their poetry in a saddening though vivid way: "the Camphor trees are dozing off," "the green hills are shivering," "the sunshine is dusty," "The rivers are covered with a layer of oil," "the birds are dejected," "stillness is sick," and "the teeth of bulldozers cut off the umbilical cord that links me to my ancestry." Like a ray of light in the darkness, Tao Yuanming has become a remote and misty illusion for these peasant workers, and *tianyuan* poetry has become reveries:

It crosses over the valley,
But never reaches our village.
Tao the hermit says so.
He is beating a tin drum,
And singing ballads of blood.
…
Chrysanthemums in September reach the den of liars,
And fish caught in the net run towards wild beasts.
Listen, the falling timbers.
Behold, the stinky moon light.
The dark light cannot light up

The numb desperate crowds of the times. (Zheng 117)

In contrast to these peasant worker poets' pessimism and desperation, what permeates the political and academic circles and the mass media is optimism about urbanization and agricultural modernization, which, in turn, are already questioned in the West. The depletion of natural resources, the decline of the groundwater level, large-scaled soil erosion, the accumulation of toxicants in the farming field, the reduction of biodiversity, air pollution, and global warming—all these begin to rock the foundation on which modern society is built. In the case of China, it hardly works if it intends to follow in Britain's and America's footsteps in its urbanization considering China's fundamental realities which include a long-established cultural tradition as large rural population.

There has been a self-evident maxim that in the course of social development, agriculture is more backward than industry, peasants are more conservative than industrial workers, and the country is less civilized than the city. It is widely believed that it is inevitable that marketization, urbanization, industrialization, and the transformation of peasants into business people and city dwellers have to be completed if a society wants to be part of the ongoing globalization. All these are established state policies in China, but their legitimacy, implementation, and value cannot be judged at this stage.

William Cowper's popular line "God made the country, and man made the town" means that the country is natural, but the town is artificial. Man should remain cautious about the orientation of human society because there might be different possibilities other than the current one. A major character of an agrarian society is man's closeness to Nature, but modern industrial society develops at the sacrifice of the country, at least in the case of China. For this reason alone, there should be a rethinking of modernization as such.

One obvious fact is that high energy-consuming urbanization runs counter to ecological nourishment and that the urban ecology is already worrying. Claude Allegre (1937–) states that "As we call for urban ecology, we should re-establish fresh rural ecology so that the urban-rural and north-south balances may be achievable" (130). Undoubtedly, without a good rural ecology, there will not be any good urban ecology. Allegre's proposal reminds people of the balance of the yin and the yang and of the principle of "knowing the bright but cleaving to the dark." However awe-inspiring, the city of concrete and iron cannot replace the village with brooks and lush vegetation, nor can dazzling city lights replace the starry or moon-lit night in the country. Urban civilization is not supposed to undo rural civilization; instead, it should rely on rural civilization for nourishment.

Committed to rural construction, Liang Shuming highly values Tao's poetry, stating that Tao's poetry is "aloof, other-worldly and transcendental," "pointing to the ontological origin," and "capable of evoking subtle holy moods, reaching inward for the deepest of human life, and elevating morality" ("SSWL" 193). The self-salvation of the country as advocated by Liang has proved to be an illusion. Many opportunities for rebuilding the countryside are gone, and new opportunities may have not come. In this context, what is important for China is that it should put its national traits in perspective and maximize the value of its cultural legacy instead of following others blindly.

Florence Eden and David Freundenberger suggest that "China should embark upon a post-modern path of agriculture" (68). For them, post-modern agriculture is not industrialized or commercialized, but eco-friendly and bioenergy-dependent; people should minimize the use of modern plowing technologies to maintain land fertility, improve sanitation, health care, and education in rural areas, and safeguard the dignity of the peasants as they respect officials and professors, maximize the support for the countryside both materially and intellectually, and bring benefits of the city to the country instead of bringing the rural population into the city (Eden et al. 68–71).

The future having presentness and pastness, industrial civilization should not sever its connection with agricultural civilization, nor should modernization and urbanization destroy the *tianyuan* tradition. If man intends to make the post-modern era a better era than the modern era, he should absorb more wisdom of life from the pre-modern era, including, of course, Tao's *tianyuan* poetry.

5.4.6 The Complexity of the "Peach-blossom Springs": An Oriental Utopia and Post-modern Romanticism

There have been many attempts to look for the location of the Peach-blossom Springs in the real world, but no consensus on the exact location has ever been reached. I visited the Donglin Temple, the Tiger Brook, Donggao Mountain, Xiechuan, the Chaisang Bridge, the Tao's intoxication rock, and Tao's tomb years ago with the assistance of a number of personalities from Jiujiang and Xingzi counties, where Tao lived. With the company of Mr. Liu Xibo, I went to the Duke Kang Valley at the foot of Mt. Lushan. The valley is deep, is small at the entrance, is covered with lush vegetation, and has steep cliffs on both sides. In it, there is a winding brook lined with peach trees on its banks. Further deep into the valley along the brook is a small hamlet with exuberant trees and bamboos, thatched cottages, and charmingly rustic flavors. According to Mr. Liu, with his former residence being not far from here, Tao must have been here and therefore wrote his "The Peach-blossom Springs" with this place as the prototype. In fact, the Peach-blossom Springs is pure imagination, though it conveys the author's emotions, ideals, and even the nation's psychological makeup. Excellent poetry is, more often than not, rooted in or associated with the unknown domains of humanity, thus transcending time and space and inspiring man. That is partly the significance of the reinterpretation of Tao Yuanming.

It may be wise for one to get immersed in Tao's "The Peach-blossom Springs." The first two paragraphs are as follows:

> In the years of Taiyuan (376–396) during the reign of Emperor Xiaowu in the Jin Dynasty, there lived in Wuling a man who made his living as a fisherman. [Rowing] up a stream one day, he forgot how far he had travelled when all at once he saw a grove of peach blossoms stretching hundreds of paces on both banks of the stream. There were no other kinds of trees but verdant grant grass in full splendor and angry flowers in profusion. Enticed by the sight, the fisherman sailed on to have a complete view of the grove.
>
> The grove stretched as far as the source of the stream, where the fisherman found a hill with an opening that seemed to be lit within. He left his boat and entered the opening. At first the opening was very narrow, barely allowing him to pass, but as he went on some dozens of paces, a wide view suddenly opened before him. There as an expansive plain scattered with orderly houses, lush fields, beautiful ponds and clumps of mulberry and bamboo trees. Footpaths crisscrossed the fields, where the crowing a cocks and the barking of dogs were heard within distance. The men and women working in the fields were attired in the same manner as the people living outside; both old and young enjoyed a happy life. (163)

Tao narrates how the Peach-blossom Springs was discovered. People living on that familiar, though strange, and secluded world were surprised at the fisherman's sudden appearance due to his intrusion and the possibility of the serene paradisiac place being discovered and disturbed. It is implied that the dwellers there were content, conservative, regressive, and unwilling to be known by the external world. Despite this, they received the "intruder" with kindness and hospitality and were "entertained" with stories about "the other world." After having heard the stories, they were surprised at the coexistence of the two worlds separated by the valley and felt lucky not to be living in the outside world and subject to all disorders. When seeing the fisherman off, they pleaded, "Don't say a word about this place to the outsiders." But the fisherman, perhaps not a man of integrity, made markings along his route and reported his discovery to the magistrate of the prefecture, who, in turn, sent men to the valley to retrace the markings. Fortunately, they got lost and failed to find the place. Liu Ziji, of worthy scholar of Nanyang, heard of the story and was keen on visiting the place, but he died of illness before he could set out. Therefore, after a brief moment of revelation, the Peach-blossom Springs vanished into nowhere.

Through Tao's account, which is vivid, the reader may feel the antithesis of the two worlds. Spatially, the Peach-blossom Springs is "other-worldly." Tao draws a clear distinction between the ideal, or more precisely idealized, world with a few words: "When the King of Qin transgressed the heavenly law, the sages left their homes and went ashore" (165). The following is Tao's description of the world somewhere in the valley:

> In the fields, each person does his very best;
> At sunset they go home and take a rest.
> Bamboos and mulberries grow in such mild clime
> While beans and crops are planted in their time.
> They raise silkworms and plough the fields in spring;
> When they reap crops, they need not pay the king.
> On busy roads, no men are seen to go,
> But dogs were heard to back and cocks to crow.
> They make sacrifices the ancient ways
> And wear the clothes they did in ancient days.
> The Children sing their songs with ringing voice;
> The grey-hair have pastime of their own choice.
> When grass grows lush, they know that spring's alive;
> When trees wither, they see autumn arrive.
> Although they do not have an almanac,
> The change of seasons helps them mark the track.
> Their lives so full of joy and bodies fit,
> They have no need to live by their wit.
> …
> How can a person from the madding crowd
> Expect to know Utopia 'neath a shroud!
> Oh that I soar to the sky on gentle breeze
> And find the men of ideal like these! (167)

What is described in the above poem is a scene of primitive farming society not touched my too much artificiality. It is the ideal society for Lao Tzu and Zhuangzi. It has been an idealized society in the minds and hearts of generations of Chinese poets and thinkers though no one is certain whether such a society ever existed in primitive times.

Temporally, time in the Peach-blossom Springs seems to be still or incessantly circular and natural. People there feel and respond to Nature physically. What governs their life, if at all, is the natural law, any violation of which will result in the harmony between man and other beings.

Tao writes, "This wonder, hidden for five hundred years, /is opened to the world as unspoiled spheres" (167). It is because people living in the Peach-blossom Springs guard their own time and space that it becomes everything that the outside world is not.

Paradoxically, in studies over the years, there has been recognition of Tao as a great poet, but, at the same time, much criticism of the most subtle, soft, and significant core of Tao's spirit. For instance, his returning to the country is seen as being degenerating, his aloofness and serenity as passiveness and self-intoxication, and his resignation of himself to Nature's transformation as surrendering to fate. Accordingly, his Peach-blossom Springs is believed by some to represent pessimism, reality dodging, and regression. These conclusions are only too natural if the critical gauges such as rationalism, the progression theory, and the Enlightenment theory are adopted.

However, in this ecological, or post-modern era, also a human era, a rethinking of Tao and his Peach-blossom Springs may lead to a more natural and reliable conclusion.

As argued earlier in this book, Tao is a classical naturalistic romanticist poet; his Peach-blossom Springs represents an oriental simple Utopian spirit, which, in fact, is classical ecological spirit. Many post-modern scholars tend to believe that there is a natural bond between post-modernism and romanticism since the former is informed and inspired by the latter in many aspects. Wang Zhihe's statement is very typical:

> The post-modernist is not entirely identical with the romanticist, but he or she has romantic genes, for he or she believes that every individual has unique value, and is irreplaceable. For the post-modernist or romanticist, it is pathetically ridiculous to follow the example of others blindly. That is why they refuse, as always, to become a part of the sweeping symphony of modern economism, materialism, money fetishism, and the grand march toward Nature, and, more than that, they choose to be discordance. Knowing the idyllic charm of Nature, they wish to live a simple natural life. Natural ecologists, they believe in what Thoreau realizes at Walden: "a man is rich in proportion to the number of things which he can afford to let alone". Letting-alone is also a form of resistance, noble resistance, since it entails great courage and wisdom. ("Preface" 25–26)

After a comparison of post-modernism and romanticism, Cheng Zhimin (1945–) finds that they converge on nature worship, the oneness of man and nature, non-standardization, national culture relativity, skepticism of science and social progress, a sense of insecurity about the reality, etc. He concludes that "Post-modernism is the reincarnated soul boy or the most fashionable expression of the long-established Western romanticism. The former inherits the tradition of the latter and carries it further forward. In a sense, post-modernism is romanticism of the twentieth century, or, extreme romanticism" (Cheng 111).

Wang Zhihe and Fan Meiyun argue that post-modernists aspire to a poetic existence, refuse to surrender to realities, value spiritual life, and want to live a simple natural life. For Wang and Fan, post-modernists are those who are full-grown personalities who can seriously taste Nature and guard the dignity of spirit in a way that is rational without being mechanical and poetic without being affected; they are closely connected with Nature and community and at the same time determined to live their own lives in a beautiful, graceful, free, and natural way. In the eyes of Wang and Fan, the role model for post-modernists is Wendell Berry (1934–), an American novelist, poet, environmental activist, cultural critic, and farmer, saying:

> Wendell Berry, for one, is a man who lives fully. As early as the early 1970s when most were subject to the trend of modernization, Berry refused to surrender to it. After he resigned from his university teaching post and his decent urban life, he settled down

as a farmer in the countryside, beginning his organic farming. He has been living there ever since, and has written more than 40 volumes of poetry, novel and essays (Wang and Fan 442).

Naturally, the reader may see a Tao Yuanming dwelling in Wendell Berry. Berry and Tao return to the countryside roughly at the same age; they both write about "returning"; they both take delight in seeing the vitality and serenity of Nature.

Can one say that post-modernism also has genes of pre-modern Chinese naturalistic romanticism? If the answer is positive, the deep reason is that there is a natural bond among Oriental Utopia, European romanticism, post-modernism, and the ecological era.

In her *The Death of Nature*, Carolyn Merchant, an American feminist eco-critic, presents an Ecotopia "in the tradition of the ancient connection assumed between nature and society" (95), and in Ecotopia, "social structure is based on an ecological philosophy of nature" (Merchant 96). In her mind, Ecotopia is like this:

> In this ecological society, reverence for trees, water, and wildlife forms the basis of an ecological religion expressed in the prayers, poetry, and little shrines of the Ecotopians. Decentralized communities, extended families, spontaneity, freedom of emotional expression, and the practice of ritual war games to deal with competitive instincts characterize the cultural norms and values. (Merchant 97)

This may be interpreted as a sign of post-modern romantics' commencement to reshape the existence of contemporary man.

In May 2010, Avenue des Champs-Elysées of Paris was decorated into a green field where the stone or concrete street was covered by trees, flowers, crops, and vegetables, and where there were pigs, dogs, and cows, thus making people in it believe, though temporarily, that they were living in a medieval village. Though it was just large-scaled "live art" supported by the government, it is intended as a reminder that the best life is "returning to nature."

In the Chinese circle of literary criticism, there have been frequent questionings like "Should man go back to the Tao-style idealized *tianyuan*?" or "Should man go back to primitive farming society in the Daoist sense?" These questionings are traditionally strong and valid. However, in the context of eco-criticism and in this ecological era, their questionings are no longer as valid or as simple.

Dedicated to finding a way out for the declining countryside, He Xuefeng (1968–) hopes that Tao Yuanming and the *tianyuan* complex he represents can help China's rural areas out of the plight, thus enabling Chinese peasants to live an easy, self-contented, and decent life. He says:

> I wish to re-establish the tianyuan life so that the peasants, after their creature comforts are secured, can enjoy clean water, green mountains, azure sky, familial bliss, neighborly love, and leisure and pleasure as Tao once enjoyed when plucking chrysanthemums. There should be labor, but no physical overdraw; there should be consumption, but no pursuit of luxury; there should be leisure, but no boredom or emptiness. In a word, they should be made happy, but their happiness is not to be gained through consumption because they cannot afford excessive consumption. They should preserve a local lifestyle which is different from consumerism and which is a low-consumption and high-wellbeing lifestyle

> emphasizing the subjective experience and human relationships. They are not necessarily rich in monetary terms, but can be rich due to their subjective experience. (He Xuefeng "Preface")

His design is essentially a Chinese edition of Ecotopia.

At the 2006 ASLA Annual Meeting & Expo and 43rd IFLA World Congress, Yu Kongjian (1963–), known in China for his ecological, post-modern, and romantic design philosophy, discussed the Peach-blossom Springs of Tao in the presence of his many international fellow landscape designers. Kong believes that the Peach-blossom Springs and the many villages that are similar to it "are the results of numerous physiological and experiential trials and adaptations over thousands of years of agricultural civilization. … It is this art of living that makes our landscape design safe, fertile, and beautiful" (26). Centered around this core philosophy, his speech has three major ideas: Namely, the Peach-blossom Springs is the origin of his landscape design as an art of existence, the disappearance of the Peach-blossom Springs brings both challenges and opportunities to landscape design, and returning to the Peach-blossom Springs should be the mission and strategy of landscape design today. He emphasizes that when the balance between man and nature is disturbed, and man's existence is in a plight, the Peach-blossom Springs must be re-established so as to restore the harmony among Heaven, the earth, man, and God or gods. Based on his decade of practice and contemplation, his *Returning to the Land* was published in 2009, which, as a symbol, reassures the deep-rooted though sometimes dormant Chinese complex of the Peach-blossom Springs and the eternal specter of Tao Yuanming dwelling in younger generation of Chinese scholars. It is a sign, if not a gospel, of resurrection not only of Tao, but also of a truly beautiful natural existence in this ecological era.

Conclusion: The Last Sacrifice and the Evocation

At the turn of the twenty-first century, there was a thought-provoking though not high-profile dialogue between a few scholars in the Chinese-language community. In his unfinished draft speech intended for the Conference on Europe, Asia and Africa over the Past Millennium, the American sinologist Benjamin Schwartz (1916–1999), who died soon after the dialogue, expresses his deep concerns over uncontrollable consumerism and materialism. After that, Yu-sheng Lin (1934–) offered an interpretation to Schwartz's speech and introduced it to Wang Yuanhua (1920–) in his 2001 Shanghai trip. After a careful reading, Wang expresses his thought as follows:

> In my humble opinion, the posthumous speech tries to convey a message that in the rapidly developing economic-technological society, man can have increasingly plentiful material comforts, thus forming consumerism and materialism. However, in such pure material pleasure and purely personal satisfaction, many people are not fully aware of the many ethical consequences of the economic-technological progress. Those die-hard followers of consumerism and materialism believe that by focusing on the economic-technological side of life, a brand-new way can be found which can eradicate the root of all sufferings in life. This is what is called by Schwartz "millenniumism." When consumerism and materialism, which originated in America, begin to spread to the rest of the world, Schwartz's speech literally throws a question to people today: "Can man, as man, live a happier life as they have more material pleasure and gratification?" I believe he may try to convey that consumerism and materialism will render the world spiritually barren. (140)

Schwartz, Yu-sheng Lin, and Wang Yuanhua share the same concern over the twenty-first century: The deprivation of spirit by consumerism and materialism, which means, in a time of affluence, man will lose his soul, whether he is in the city or in the country. An opportunity, if not a turning point, is that traditional Western humanism is trying to form an alliance with Nature, and therefore "reveres eco-movements" (149). Schwartz believes that in Western literature, the Shakespearean narrative does not suffice any longer. Should people today also draw inspiration from the Chinese tradition of Nature and man? Tao Yuanming's philosophy of "knowing the bright but cleaving to the dark," his abandon, aloofness, and serenity, his romantic resignation to Nature's transformation, his returning to Nature, his pure and simple *tianyuan* complex, and his honest poverty, ease

S. Lu, *The Ecological Era and Classical Chinese Naturalism*, China Academic Library, DOI 10.1007/978-981-10-1784-1

and idling—could all these inform and inspire people today who seem to go in an opposite direction of Tao?

In this ecological era when the issue of "Nature and man" is not a meta-question, but the meta-question of humanity, people have to consider where they are going. In fact, the unknown poets in *The Book of Poetry*, Qu Yuan, Tao Yuanming, Wang Wei, Li Bai, Du Fu, Bai Juyi, Su Shi, Lu You, Xin Qiji, Tang Xianzu, Cao Xueqin, Shen Congwen, Wang Zengqi, Wei An, Hai Zi, and to today's migrant worker poets—they all go along a road with an intersection of Nature and man. Tao Yuanming, the poet of poets, founder of the *tianyuan* school, avatar of Nature, and epitome of poetic dwelling, remains an eternal soul in the history of Chinese literature.

Tao Yuanming was a prophet of the meta-question of Nature and man. His writings convey his natural philosophy whose core is "knowing the bright but cleaving to the dark," which, in turn, has influenced Heidegger's philosophy.

As an oriental naturalist romantic poet, Tao has a number of Western naturalistic romanticist soulmates, including Rousseau and Thoreau.

Tao's reception in China has been changing with the fundamental realities and ethos of the times. His second death in contemporary China, following his true death about 1600 years ago, symbolizes the decline of his poetry, of the pursuit of spiritual well-being, of poetic dwelling, and of human existence today.

However, according to Derrida, a specter never dies; it is also the nostalgic waiting for redemption, then people today have every good reason to wait for Tao's tender ethereal specter's breaking of the cage built with iron and concrete, rubber, institutions, consumerism, and materialism. Tao's specter may help reveal the truthfulness and naturalness of Nature and evoke a truly beautiful existence. The many consequences of modernization urge the world to think about returning to the true, the simple and the natural, which is at the core of pre-modern oriental wisdom of life. In this post-modern context, the image of Tao Yuanming and his soulmates such as Rousseau and Thoreau loom large.

Lao Tzu says, "Now *ta* (great) also means passing on, and passing on means going Far Away, and going far away means returning." (53). When modernism and post-modernism go to extremes, there should be, and possibly will be, some form of return in the same way a caterpillar returns after it reaches the end of the twig for its own sake.

Will Tao be resurrected at all? Now that the crises of contemporary times, ecological, spiritual, or otherwise, are escalating, man, who is being driven into desperation, may be forced to "return" or "regress" before he is past salvation. Though agricultural civilization is slipping into oblivion, yet it, along with the beautiful dwelling it nurtures and nourishes, still has tremendous appeal, at least for the specter of Tao and for those who are aware of the significance and returning. We are waiting for the return of Tao's specter.

It may not be inappropriate to cite a stanza of Qu Yuan to serve as the finale:

> I gaze into infinity,
> My heart aching at the spring scene.
> Oh, Soul! Return! The south's imbu'd
> With a yearning for you so keen! (Zhuo 197).

References

Allegre, C. (2003). *Urban ecology and rural ecology*. Beijing: The Commercial Press.
Baudrillard, J. (2000). *The consumer society: Myths and structures*. Nanjing: Nanjing University Press.
Berlin, I. (1999). In H. Hardy (Ed.), *The roots of romanticism*. New Jersey: Princeton University Press.
Bertalanffy, L. (1981). In P. A. LaViolette (Ed.), *A systems view of man*. Boulder: Westview Press.
Brandes, G. (1902). *Main currents in nineteenth century literature*. London: William Heinemann.
Burroughs, J. (1877). *Birds and poets: With other papers*. Cambridge: Cambridge University Press.
Bury, J. (1920). *The idea of progress*. London: Macmillan.
Cassirer, E. (1954). *The question of Jean-Jacques Rousseau*. New York: Columbia University Press.
Chan, W. (Ed.). (2006). *Selected Chinese philosophical literature*. Nanjing: Jiangsu Education Press.
Chen, D. (1987). *Works of Duxiu*. Hefei: Anhui People's Press.
Chen, G. (1985). *Lao Tzu: Footnotes and criticism*. Beijing: China Bookstore.
Chen, G. (2007). *Zhuangzi: New notes and interpretation*. Beijing: The Commercial Press.
Chen, J. (2009). *Notes on reading the works of Mao Zedong*. Beijing: Sanlian Bookstore.
Chen, X. (1997). Tao Yuanming writing an elegy. In *New Chinese Literature 1949–1976 (II)*.
Chen, X. (2007). Forgetting and evocation: The modern tradition and contemporary writers. *Review of Contemporary Writers, 6*.
Chen, X. (2009). *The bottom line of Derrida*. Beijing: Peking University Press.
Chen, Y. (2008). *Sinology in Britain: Phase characteristics and reasons* (Vol. 107). Taiwan: International Sinology.
Chen, Y. (2009). *A compilation of the Jinming study manuscripts*. Shanghai: Sanlian Bookstore.
Cheng, Z. (2005). *A survey of post-modern philosophical ethos*. Beijing: Huaxia Hanlin Press.
Collingwood, R. G. (1945). *The idea of nature*. Oxford: Oxford University Press.
Dai, Y. (2002). *The power of literary history*. Beijing: Peking University Press.
Derrida, J. (2006). *Specters of Marx*. Trans. Peggy Kamuf. New York: Routledge.
Durning, A. (1992). *How much is enough? The consumer society and the future of the earth*. New York: Norton.
Eden, F., Freundenberger, D., & Zhou, B. (2009). China should embark upon a postmodern path of agriculture. *Guangzhou: Modern Philosophy, 1*.
Fang, D. (1992). *Ideals of life and cultural types*. Shenzhen: China Radio and Television Press.
Feng, C. (2006). *The spirit of Jung*. Haikou: Hainan Publishing House.

S. Lu, *The Ecological Era and Classical Chinese Naturalism*,
China Academic Library, DOI 10.1007/978-981-10-1784-1

Feng, Q. (2001). *A dictionary of philosophy (I)*. Shanghai: Shanghai Dictionary Press.
Friedlander, E. (2004). *J.J. Rousseau: An afterlife of words*. Cambridge: Harvard University Press.
Fung, Y. (2001). *A complete collection of the three-pine study (V)*. Zhengzhou: Henan People's Press.
Fung, Y. (2007). *A short history of Chinese philosophy*. Tianjin: Tianjin Social Sciences Press.
Gao, H. (1984). *Notes to the book of change*. Beijing: The Chinese Bookstore.
Giddens, A. (1991). *The consequences of modernity*. Stanford: Stanford University Press.
Giddens, A. (1994). *Beyond left and right: The future of radical politics*. Oxford: Polity Press.
Glen, A. L. (1996). Revaluing nature: Toward an ecological criticism in Cherylly and Harold from meds. In *The ecocriticism reader: Landmarks in ecology*. Atlanta: The University of Georgia Press.
Gong, P. (2005). *The ethos of the Han Dynasty*. Beijing: The Commercial Press.
Gore, S. A. (2007). *Earth in the balance: Forging a new common purpose*. New York: Earthscan.
Gu, Y. (1997). *Rizhao Lu*. Lanzhou: Gansu Minzu Publishing House.
Guo, J., et al. (2003). *On Walter Benjamin*. Changchun: Jilin People's Press.
He, L. (1999). *Culture and life*. Shanghai: The Commercial Press.
He, X. (2007). *The way out for the country*. Jinan: Shandong People's Publishing House.
He, Y. (1962). *Remarks on poetry. The reprinted version of the Daoguang's Reign.in A Collection of Resources on Tao Yuanming (I)*. Beijing: China Bookstore.
Heidegger, M. (2002). *Elucidations of Hölderlin's poetry*. Beijing: The Commercial Press.
Hou, W., et al. (1980). *A general history of Chinese thought (Volume I)*. Beijing: The People's Press.
Hsu, C. (2010). *The analects of Hsu Cho-yun*. Nanchang: Guangxi Normal University Press.
Hu, B. (2007). *Notes on my reading of Tao Yuanming*. Shanghai: East Normal University Press.
Hu, L. (2004). *A history of Chinese literature*. Shanghai: Shanghai Social Sciences Press.
Hu, S. (2006). *A history of vernacular literature*. Hefei: Anhui Education Press.
Huret, J. (1891). *Paul Gauguin discussing his paintings*. Paris: L'Écho de Paris.
Jay, M. (1996). *The dialectical imagination: A history of the Frankfurt School and the Institute of Social Research, 1923–50*. Guangzhou: Guangdong People's Press.
Jin, Y. (2013a). *Tao, nature, and man*. Beijing: The People's Press.
Jin, Y. (2013b). *The complete works of Jin Yuelin*. Beijing: The People's Publishing House.
Joyce, J. (2000). In K. Barry (Ed.), *Occasional, critical and political writings*. London: Penguin.
Jung, C. (1997). *Selected works of Jung: Returning to the spiritual homeland*. Beijing: The Reform Press.
Kelly, C. (1987). *Rousseau's exemplary life: The confessions as political philosophy*. Ithaca and London: Cornell University Press.
Kuang, J. (2002). *Refined documents about Korean poetic remarks on Chinese poetry*. Beijing: The Chinese Press.
Kubin, W. (2008). *The history of twentieth century Chinese literature*. Shanghai: East China Normal University Press.
Lao T. (1999). *Tao Te Ching*. Trans. Arthur Waley. Beijing: Foreign Language Teaching and Research Press.
Li, C. (2007). *A biographical study of Tao Yuanming*. Tianjin: Tianjin People's Press.
Li, P. (2004). *The oriental literary art in the western eyes*. Shanghai: Shanghai Education Press.
Li, Y. (1995). *Notes on my idle life*. Beijing: Writers' Press.
Li, Z. (1982). *The evolution of aesthetics*. Beijing: China Social Sciences Press.
Liang, Q. (1924). *Tao Yuanming*. Shanghai: The Commercial Press.
Liang, S. (1992). *Self-selected works of Liang Shuming*. Beijing: Beijing Normal University Press.
Liang, Z. (2005). *The poetic and the picturesque*. Beijing: The Central Compilation Press.
Lin, G. (2005). *A history of Chinese literature*. Xiamen: Lujiang Publishing House.
Liu, B. (2001). *A portrait of Foucault's thought*. Shanghai: Shanghai People's Press.

Liu, D. J. (2007). *A history of Chinese literature* (p. 140). Tianjin: Baihua Literature and Art Publishing House.
Liu, X. (2007). *Poetic philosophy*. Shanghai: East China Normal University.
Liu, Z. (2006). *The reception of Tao Yuanming in the Tang dynasty*. Beijing: China Social Sciences Press.
Lu, X. (1981). *Complete works of Lu Xun (III)*. Beijing: People's Literature Press.
Luo, D. (1994). *A rare meeting with Henri Michau* (Vol. 1). Shanghai: Foreign Literature Studies.
Luo, Z. (2005). *Metaphysics and the mindset of Wei and Jin scholars*. Tianjin: Tianjin Education Press.
Luo, Z. (2006). *A history of literary thought: From 220 to 589*. Shanghai: Chung Hwa Book Co.
Lyotard, J. F. (2000). Ecology as discourse of the secluded. In L. Coupe (Ed.), *The green studies reader*. London: Routledge.
Marcuse, H. (2007). In A. Feenberg & W. Leiss (Eds.), *The essential Marcuse: Selected writings of philosopher and social critic Herbert Marcuse*. Boston: Beacon Press, .
Marx, K. (1985). *Economic and philosophical manuscripts of 1844*. Beijing: The People's Press.
McKibben, B. (2003). *The end of nature*. London: Bloomsbury Publishing.
Merchant, C. (1989). *The death of nature: Women, ecology, and the scientific revolution*. New York: Harper & Row.
Morin, E. (1999). *The lost paradigm: A study of human nature*. Beijing: Peking University Press.
Moscovici, S. (2005). *De la Nature*. Shanghai: Sanlian Bookstore.
Mou, Z. (1984). *Historical philosophy*. Taipei: Students' Bookstore (Taiwan).
Muir, J. (1909). *Our national parks*. Boston: Houghton Mifflin Company.
Needham, J. (1991). *Science and civilization in China* (Vol. 2). Cambridge: Cambridge University Press.
Nietzsche, F. (1968). In W. Kaufmann (Ed.), *The will to power*. New York: Random House.
Odum, E., & Barrett, G. W. (2005). *Fundamentals of ecology* (5 ed.). California: Thomson Brooks/Cole.
Poggeler, O. (1987). In G. Parkes (Ed.), *West-East dialogue: Heidegger and Lao-Tzu*. Heidegger and Asian Though. University of Hawaii.
Priest, G. (2003). Where is philosophy at the start of the twenty-first century? In *Proceedings of the Aristotelian Society*. New Jersey: Wiley, New Series 103.
Prigogine, I. (1987). *From Chaos to order: A new dialogue between man and nature*. Shanghai: Shanghai Translation Press.
Qian, M. (2002). *A general study of Lao Tzu and Zhuangzi* (p. 368). Shanghai: Sanlian Bookstore.
Qu, Y. (2006) *The verse of Chu*. Trans. Zhuo Zhenying. Changsha: Hunan People's Publishing House.
Radkau, J. (2004). *Nature and power*. Shijiazhuang: Hebei University Press.
Rescher, N. (1998). *Complexity: A philosophical overview*. New Brunswick: Transaction Publishers.
Rilke, R. M. (1996). *The book of pilgrimage*. London: Penguin Books.
Rosen, D. (1996). *The Tao of Jung*. New York: Viking.
Rousseau, J.-J. (1979). *Emile: Or, on education*. Trans. Bloom, Allan. New York: Basic Books.
Rousseau, J.-J. (1992). *The reveries of the solitary walker*. Indianapolis/Cambridge: Hackett Publishing Company.
Rousseau, J.-J. (1995). *The confessions and correspondence*. Trans. Kelly, Christopher, Kelly. Hanover: The University Press of New England.
Rousseau, J.-J. (1999). *Discourse on Political Economy and the Social Contract*. Trans. Christopher Betts. Oxford: Oxford University Press.
Sainte-Beuve, C. (1905). *Portraits of the eighteenth century*. Trans. Wormeley, Katharine. New York: G.P. Putnam's Sons.
Sayre, R. (1996). *Henry David Thoreau: Writings (II)*. Shanghai: Sanlian Bookstore.
Scheler, M. (1999). *Selected works of Max Scheler*. Shanghai: Sanlian Bookstore.

Scheler, M. (2002). Man's place in nature. In M. W. Dennis (Ed.), *Interpreting Man*. The Davies Group.
Schopenhauer, A. (1987). *The world as will and representation*. Beijing: The Commercial Press.
Schweitzer, A. (1946). *Civilization and ethics* (3rd ed.). London: Adam & Charles Black.
Schweitzer, A. (1987). *The philosophy of civilization*. New York: Prometheus Books.
Seubold, G. (1993). *Heidegger's analysis of technology of the new age (Heidegger's Analyse der neuzeitlichen Technik)*. Beijing: China Social Sciences Press.
Shang, J. (2006). *A mock dialogue with later Derrida*. Shanghai: Tongji University Press.
Shang, J. (2008). *Shang Jie on Rousseau*. Beijing: Peking University Press.
Shen, D. (2013). *Remarks on poetry*. Beijing: People's Literature Press.
Shigeru, O. (2002). *A new reading of Tao Yuanming and Li Bai*. Shanghai: Shanghai Classics Press.
Sim, S. (1999). *Derrida and the end of history*. Cambridge: Icon Books.
Slovic, S. (2010). *Going away to think: Engagement, retreat, and ecocritical responsibility*. Beijing: Peking University Press.
Stiegler, B. (1998). *Technics and time: The fault of epimetheus*. Stanford: Stanford University Press.
Sun, X. (Ed.). (1996). *Selected works of Heidegger (I)*. Shanghai: Sanlian Bookstore.
Tan, S. (1954). *Complete works of Tan Sitong (III)*. Shanghai: Sanlian Bookstore.
Tang, Y. (2001). *Essays on the metaphysics of the Wei and the Jin dynasties*. Shanghai: Shanghai Classics Publishing House.
Tao, Y. (2003). *The complete works of Tao Yuanming*. Trans. Wang Rongpei. Changshu: Hunan People's Publishing House.
Thoreau, H. D. (1997). *Walden*. Changchun: Jilin People's Press.
Tu, A. (2009). Preface to the lake poetic souls. Yang, Deyu, Trans. In *Selected poems of Wordsworth*. Liuzhou: Guangxi Normal University Press.
Tu, W. (2004). *The continuity of being: The chinese versions of nature* (Vol. 1, pp. 88–93). Beijing: World Philosophy.
Tu, W. (2005). *Dialogue and innovation*. Guilin: Guangxi Normal University Press.
Wang, F. (1964). *Interpretation of Zhuangzi*. Shanghai: Chung Hwa Book Co.
Wang, X. (2010). Tao Yuanming's ecological dimensions. In S. Yanwei (Ed.), *Selected eco-critical essays*. Hangzhou: Zhejiang Gongshang University Press.
Wang, Y. (2004). *A collection of the latest writings of the Qingyuan study*. Shanghai: Wenhui Press.
Wang, Z. (2005). Preface to the Chinese Edition. In S. Sim (Ed.), *Derrida and the end of history*. Beijing: Beijing University Press.
Wang, Z., & Fan, M. (2011). *The second enlightenment* (p. 442). Beijing: Peking University Press.
Weber, M. (2001). *The protestant ethic and the spirit of capitalism*. New York: Routledge.
Wei, A. (2000). *After the sunrise*. Beijing: The Workers' Publishing House.
Whitman, W. (2001). *Leaves of grass*. Beijing: Jiuzhou Press.
Ye, S. (1992). *Chinese philosophy of mythology*. Beijing: China Social Sciences Press.
Ye, X., & Wang, S. (Eds.). (2006). *A history of western philosophy (V)*. Nanjing: Jiangsu People's Press.
Yu, Y. (2004). *Modern confucianism: Retrospect and prospect*. Beijing: Sanlian Bookstore.
Yuan, X. (2009). *A study of Tao Yuanming*. Beijing: Peking University Press.
Zang, K., & Zhou, Z. (1990). *A reading of the poetry of Mao Zedong*. Beijing: China Youth Press.
Zeng, F. (2007). *Chinese aesthetics at a time of transformation*. Beijing: The Commercial Press.
Zhang, S. (2007). *The interaction between heaven and man: Dilemmas and choices of eastern and western philosophies*. Beijing: The People's Press.
Zhang, X. (2001). *From phenomenology to confucius*. Beijing: The Commercial Press.
Zhang, Z. (2007). Dao: The Order and the Sentence Pattern: A Reading of Chapter 28 of Lao Tzu. *Jiangsu Social Sciences*, *6*.
Zheng, X. (2008). *Selected poems of Zheng Xiaoqiong*. Guangzhou: Huacheng Publishing House.

Zhao, Y. (2007). *From Hussel to Derrida: Lectures on western literary theory*. Beijing: Sanlian Bookstore.

Zhu, G. (1982). *A collection of essays of Zhu Quangqian on aesthetics (II)*. Shanghai: Shanghai Literature and Arts Press.

Zhu, Q. (2010). *On poetry*. Changsha: Yuelu Bookstore.

Zhu, Z. (2009). *A collection of essays on classical Chinese literature (II)*. Shanghai: Shanghai Classics Publishing House.

Zhuangzi. (1999). *Zhuangzi*. Trans. Wang Rongpei. Changsha: Human People's Publishing House.

Zong, B. (1981). *Strolling in aesthetics*. Shanghai: Shanghai People's Press.

Габитова, P. M. (2007). *German romantic philosophy*. Beijing: China Central Compilation Bureau Press.

MIX
Papier aus verantwortungsvollen Quellen
Paper from responsible sources
FSC® C105338
FSC
www.fsc.org

If you have any concerns about our products,
you can contact us on
ProductSafety@springernature.com

In case Publisher is established outside the EU,
the EU authorized representative is:
Springer Nature Customer Service Center GmbH
Europaplatz 3, 69115 Heidelberg, Germany

Printed by Libri Plureos GmbH
in Hamburg, Germany